SpringerBriefs in Electrical and Computer Engineering

Control, Automation and Robotics

More information about this series at http://www.springer.com/series/10198

Yunfei Xu · Jongeun Choi · Sarat Dass
Tapabrata Maiti

Bayesian Prediction and Adaptive Sampling Algorithms for Mobile Sensor Networks

Online Environmental Field Reconstruction in Space and Time

Springer

Yunfei Xu
Michigan State University
East Lansing, MI
USA

Jongeun Choi
Michigan State University
East Lansing, MI
USA

Sarat Dass
Department of Statistics
Michigan State University
East Lansing, MI
USA

Tapabrata Maiti
Department of Statistics
Michigan State University
East Lansing, MI
USA

ISSN 2191-8112 ISSN 2191-8120 (electronic)
SpringerBriefs in Electrical and Computer Engineering
ISSN 2192-6786 ISSN 2192-6794 (electronic)
SpringerBriefs in Control, Automation and Robotics
ISBN 978-3-319-21920-2 ISBN 978-3-319-21921-9 (eBook)
DOI 10.1007/978-3-319-21921-9

Library of Congress Control Number: 2015950872

Springer Cham Heidelberg New York Dordrecht London

Printed on acid-free paper

Springer International Publishing AG Switzerland is part of Springer Science+Business Media (www.springer.com)

To our loving parents and beautiful families

Preface

We have witnessed a surge of applications using static or mobile sensor networks interacting with uncertain environments. To treat a variety of useful tasks such as environmental monitoring, adaptive sampling, surveillance, and exploration, this book introduces a class of problems and efficient spatio-temporal models when scalar fields need to be predicted from noisy observations collected by mobile sensor networks. The book discusses how to make inference from the observations based on the proposed models and also explores adaptive sampling algorithms for robotic sensors to maximize the prediction quality subject to constraints on memory, communication, and mobility.

The objective of the book is to provide step-by-step progress in chapters for readers to gain better understanding of the interplay between all the essential constituents such as resource-limited mobile sensor networks, spatio-temporal models, data-driven prediction, prediction uncertainty, and adaptive sampling for making better predictions. The book builds on previous collective works by the authors and is not meant to provide a comprehensive review of the topics of interest. Specifically, materials from the previous publications by the authors [1–5] make up a large portion of the book.

In this book, a spatio-temporal scalar field is used to represent the collection of scalar quantities of interest, such as chemical concentration or biomass of algal blooms (e.g., see Fig. 1.3), transported via physical processes. To deal with complexity and practicality, phenomenological and statistical modeling techniques are used to make inference from noisy observations collected, taking into account a large scope of uncertainties. To this end, nonparametric models such as Gaussian processes and Gaussian Markov random fields (GMRFs), along with their prediction and adaptive sampling algorithms, will be explored and tailored to our needs. The importance of selecting a Gaussian process prior via hyperparameters for given experimental observations is illustrated (Chap. 3). Adaptive sampling to improve the quality of hyperparameters is proposed (Chap. 3). Memory efficient prediction based on truncated observations in space and time as well as the collective mobility based on distributed navigation are discussed (Chap. 4). While the book starts with a rather simple empirical Bayes approach (Chap. 3), as we move through further chapters, we discuss recent efforts with a fully Bayesian perspective to maximize the flexibility of the models under various uncertainties while minimizing the computational complexity (Chaps. 5 and 7). A fully Bayesian framework is adopted here as it offers several advantages when inferring parameters and processes from highly complex models (Chaps. 5 and 7). The Bayesian approach requires prior distributions to be elicited for model parameters that are of interest. Once the priors are elicited, the Bayesian framework is flexible and effective in incorporating all uncertainties as well as information (limited or otherwise from data) into a single entity, namely, the posterior. The fully Bayesian approach thus allows additional sources and extent of uncertainties to be integrated into the inferential framework, with the posterior distribution effectively capturing all aspects of uncertainties involved. Subsequently, the practitioner needs only to focus on different components of the posterior to obtain inference separately for the parameters of interest, nuisance parameters, and hyperparameters. The fully Bayesian approach also allows data to select the most appropriate values for nuisance parameters and hyperparameters automatically and achieve optimal inference and prediction for the scalar field. In this book, a fully Bayesian approach for spatio-temporal Gaussian process regression will be formulated for resource-constrained robotic sensors to fuse multifactorial effects of observations, measurement noise, and prior distributions for obtaining the predictive distribution of a scalar environmental field of interest. Traditional Markov Chain Monte Carlo (MCMC) methods cannot be implemented on resource-constrained mobile sensor networks due to high computational complexity. To deal with complexity, the Bayesian spatio-temporal models will be carefully tailored (Chap. 5). For example, we will approximate a Gaussian process with a GMRF for computational efficiency (Chaps. 6 and 7). A new spatial model is proposed via a GMRF (Chap. 6). In addition, ways to improve computational efficiency will be proposed in form of empirical Bayes and approximate Bayes instead of MCMC-based computation. For some special cases, the developed centralized algorithms will be further refined in a distributed manner such that mobile robots can implement distributed algorithms only using local information available from neighboring robots over a proximity communication graph (Chaps. 4–6).

We note that although regression problems for sensor networks under location uncertainty have practical importance, they are not considered in this book. The interested reader is referred to [6, 7] (centralized scheme) and [8] (distributed scheme) for further information on this topic.

Organization

This book is organized as follows: Chapter 1 gives some background information and a summary for each chapter. In Chap. 2, we introduce the basic mathematical notation that will be used throughout the book. We then describe the general

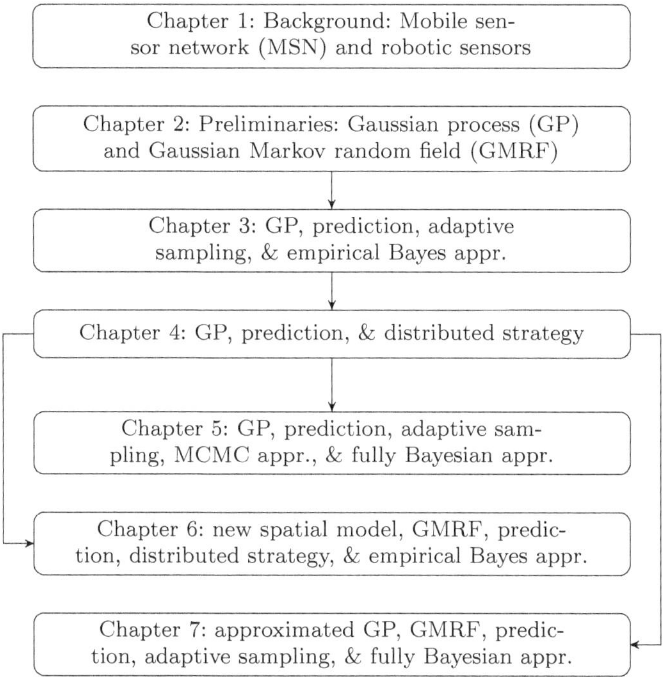

Fig. 1 Organization of chapters along with keywords

Gaussian process and its usage in nonparametric regression problems. The notations for mobile sensor networks are also introduced in Chap. 2. In Chap. 3, we deal with the case where hyperparameters in the covariance function is deterministic but unknown. We design an optimal sampling strategy to improve the maximum likelihood estimation of these hyperparameters. In Chap. 4, we assume the hyperparameters in the covariance function are given; they can be obtained using the approach proposed in Chap. 3. We then analyze the error bounds of prediction error using Gaussian process regression with truncated observations. Inspired by the error analysis, we propose both centralized and distributed navigation strategies for mobile sensor networks to move in order to reduce prediction error variances at points of interest. In Chap. 5, we consider a fully Bayesian approach for Gaussian process regression in which the hyperparameters are treated as random variables. Using discrete prior probabilities and compactly supported kernels, we provide a way to design sequential Bayesian prediction algorithms that can be computed in constant time as the number of observations increases. To cope with the computational complexity brought by using standard Gaussian processes with covariance functions, in Chap. 6, we exploit the sparsity of the precision matrix by using Gaussian Markov random fields (GMRFs). We first introduce a new class of Gaussian processes with built-in GMRF and show its capability of representing a wide range of nonstationary physical processes. We then derive the formulas for predictive statistics and design sequential prediction algorithms with fixed complexity. In Chap. 7, we consider a discretized spatial field that is modeled by a GMRF with unknown hyperparameters. From a Bayesian perspective, we design a sequential prediction algorithm to exactly compute the predictive inference of the random field. An adaptive sampling strategy is also designed for mobile sensing agents to find the most informative locations in taking future measurements in order to minimize the prediction error and the uncertainty in the estimated hyperparameters simultaneously.

Keywords for chapters are summarized in Fig. 1. While each chapter is self-contained and so can be read independently, arrows in Fig. 1 recommend possible reading sequences for readers.

Acknowledgments

We would like to thank Songhwai Oh at Seoul National University for his suggestions and contribution to Chap. 4. We also thank the National Science Foundation RET teacher, Alexander Robinson, undergraduate student David York, and Ph.D. student Huan N. Do at Michigan State University for collecting the experimental data using a robotic boat in Chap. 3. We thank Jeffrey W. Laut, Maurizio Porfiri (NYU Polytechnic School of Engineering), Xiaobo Tan (Michigan State University), and Derek A. Paley (University of Maryland) for providing pictures of their robots used in the introduction of Chap. 1.

The authors Yunfei Xu and Jongeun Choi have been supported in part by the National Science Foundation through CAREER Award CMMI-0846547. This support is gratefully acknowledged. Any opinions, findings, and conclusions, or recommendations expressed in this book are those of the authors and do not necessarily reflect the views of the National Science Foundation.

Santa Clara, California — Yunfei Xu
East Lansing, Michigan — Jongeun Choi
Seri Iskandar, Malaysia — Sarat Dass
East Lansing, Michigan — Tapabrata Maiti
May 2015

Contents

Chapter 1
Introduction

1.1 Background

Sensor networks are ubiquitous due to the recent technological breakthroughs in micro-electro-mechanical systems (MEMS), wireless communications, and embedded systems [9, 10]. A sensor network consists of a collection of low-cost, low-power, and multifunctional sensing devices that communicate over finite distances. A flexible and application-specific operating system, TinyOS, was developed at UC Berkeley for sensor networks with severe memory and power constraints [11]. TinyOS runs on small and cheap wireless sensor nodes (e.g., MICA2DOT from Crossbow Technology, Inc., CA, USA) as shown in Fig. 1.1a. Such sensor nodes have been equipped with various environmental and ambient sensors such as temperature sensors, lighting sensors, chemical sensors, accelerometers, and RFID readers along with the communication capability with neighbors via low-power wireless communication to form a wireless ad hoc sensor network with up to 100,000 nodes [10].

Endowing the nodes in a sensor network with mobility significantly increases the sensor network's sampling capabilities [12, 13]. The sensor networks which consist of mobile sensing agents are more flexible than the ones with only static nodes. A conceptual picture of a distributed mobile sensor network with a (R-disk) proximity communication graph model is shown in Fig. 1.1b, which assumes that a robotic sensor can communicate with its neighboring robots within distance R. Devised in the Laboratory of Intelligent Systems at the Swiss Federal Institute of Technology, the Swarming Micro Air Vehicle Network (SmavNet) depicted in Fig. 1.1c allows a single operator to control an entire swarm of cheap unmanned aerial vehicles (UAVs) for search and rescue operation [14]. 3D Robotics, Inc., Berkeley, USA produces "personal drones" such as DIY drone kits as well as ready-to-fly quadrotors, multirotors, and fixed wing UAVs based on open source UAV autopilot platforms. A quadrotor with a camera from 3D Robotics is shown in Fig. 1.2.

Biologists and other scientists are interested in leveraging recent technological advances [10, 15, 16] by deploying mobile sensor networks for environmental and

Y. Xu et al., *Bayesian Prediction and Adaptive Sampling Algorithms for Mobile Sensor Networks*, SpringerBriefs in Control, Automation and Robotics, DOI 10.1007/978-3-319-21921-9_1

Fig. 1.1 **a** Wireless microsensor mote (MICA2DOT) from Crossbow Technology, Inc., CA, USA, (www.xbow.com). **b** Distributed mobile sensor network with a (R-disk) proximity communication graph for environmental monitoring (credit: Justin Mrkva). **c** Swarming Micro Air Vehicle Network (SmavNet) developed in the Laboratory of Intelligent Systems at the Swiss Federal Institute of Technology for search and rescue operation (Photo courtesy of Laboratory of Intelligent Systems at EPFL, http://lis.epfl.ch)

wildlife monitoring. For example, one of the pressing societal concerns about water quality is the proliferation of harmful algal blooms in ponds, lakes, rivers, and coastal ocean worldwide. A satellite image of a 2011 significant harmful algal bloom in western Lake Erie is shown in Fig. 1.3. The excessive growth of cyanobacteria leads to a decaying biomass and oxygen depletion, which are detrimental to fish and other aquatic life as well as to land animals and humans consequently (due to the produced

Fig. 1.2 Personal drone (IRIS+) manufactured by 3D Robotics Inc., Berkeley, USA (credit: 3D Robotics, http://3drobotics.com)

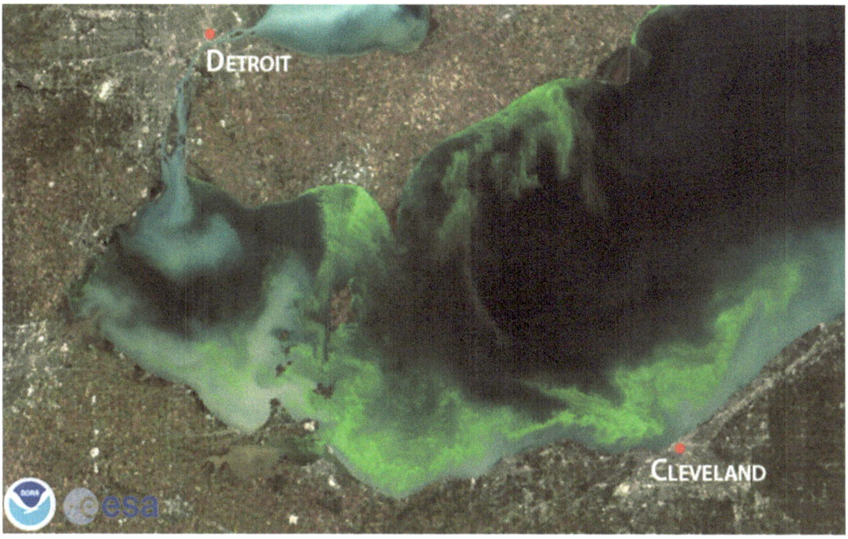

Fig. 1.3 Satellite image of 2011 significant harmful algal bloom in western Lake Erie in Michigan, which impacted over half of the lake shore (credit: MERIS/ESA, processed by NOAA/NOS/NCCOS, http://www.noaanews.noaa.gov)

toxins that deteriorate water quality) [17–19]. Deploying mobile sensor networks can be a viable way to reconstruct and monitor such harmful algal blooms [20, 21].

Indeed, we have seen the increasing exploration of robotic technologies in aquatic sensing [20, 22–25]. A robotic boat was used in concert with stationary buoys to form an aquatic microbial system [20]; spatiotemporal aquatic field reconstruction was implemented using inexpensive, low-power, robotic fish in [21] (see also Fig. 1.5a for gliding robotic fish [26]); and low-cost, self-sustained mobile surface vehicles have been designed for environmental monitoring as part of the citizen science project Brooklyn Atlantis [23] (see Fig. 1.4). A robotic boat equipped with a depth sensor, as

(a)

(b)

Fig. 1.4 Robots developed by NYU Polytechnic School of Engineering for image and water quality data collection as part of the citizen science project Brooklyn Atlantis (http://www.brooklynatlantis.poly.edu) [23] (credit: Jeffrey W Laut)

shown in Fig. 1.5b, can sample the depth of a lake for its estimation [27]. Autonomous underwater vehicles (AUVs) are being developed as an important tool in oceanography, marine biology, and other maritime applications [28–30]. Autonomous sea gliders are another noteworthy example. These battery-powered, buoyancy-driven vehicles can travel thousands of miles horizontally, for many months, without changing or recharging batteries [31–34]. With the networks of gliders as shown in Fig. 1.6, adaptive sampling has been demonstrated in Monterey Bay, California [35–37].

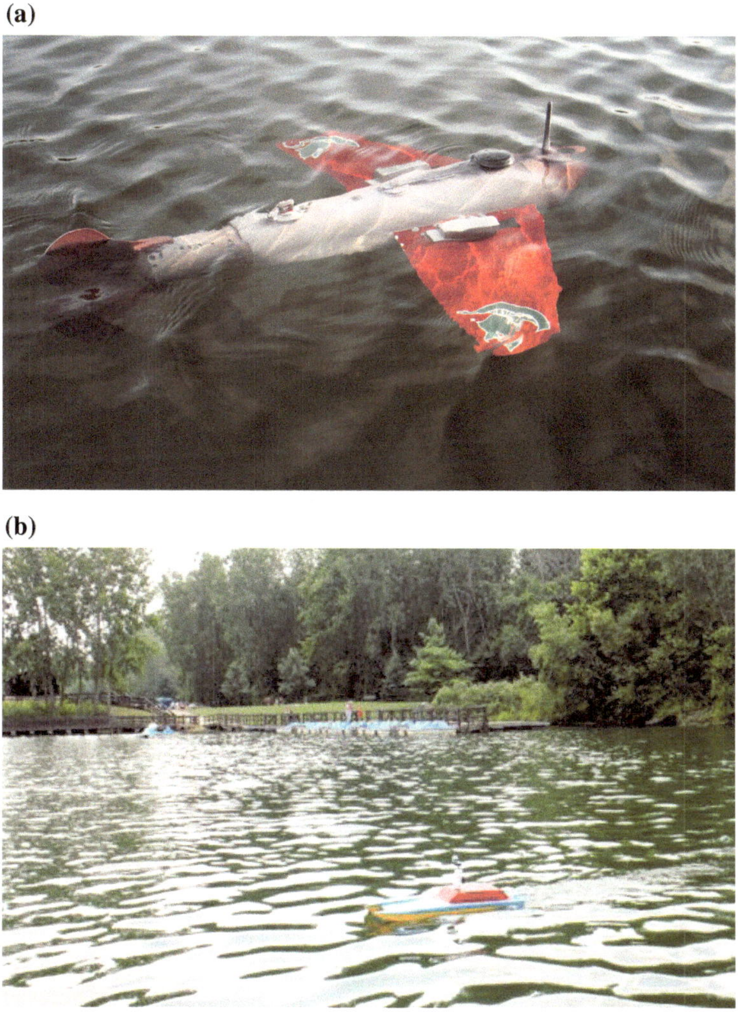

Fig. 1.5 **a** Gliding robotic fish "Grace" sampling harmful algae in the Wintergreen Lake, Michigan [26] (credit: Xiaobo Tan). **b** Robotic boat sampling depth near Hawk Island, Lansing, Michigan (credit: Jongeun Choi)

The robotic sensor technologies have then brought an increasing exploitation of navigation of mobile sensor networks and robotic sensors interacting with uncertain environments [2, 35–42]. A necessity in such scenarios is to design algorithms to process collected observations from environments (e.g., distributed estimators) for robots such that either the local information about the environment can be used for local control actions or the global information can be estimated asymptotically.

The approach of designing such algorithms takes two different paths depending on whether it uses an environmental model in space and time or not. Without

Fig. 1.6 Gliders used for adaptive sampling [35–37] (credit: Derek A. Paley)

environmental models, extremum seeking control has been proven to be very effective for finding a source of a signal (chemical, electromagnetic, etc.) [38, 39]. Distributed algorithms for stochastic source seeking with mobile robot networks have been developed for both cases with and without the mutual information model between their expected measurements and the expected source location [43]. A unifying framework of distributed stochastic gradient algorithms that can deal with coverage control, spatial partitioning, and dynamic vehicle routing problems in the absence of a priori knowledge of the event location distribution has been presented in [40].

A drawback of the spatial model free approach is that it limits its task to finding the maximum (or minimum) point of the environmental field. To tackle a variety of useful tasks such as the exploration, estimation, prediction, and maximum seeking of a scalar field, it is essential for robots to have a spatial (and temporal) field model [2–4, 36, 41, 42, 44–50]. Although control algorithms for mobile robots have been developed based on computationally demanding, physics-based field models [51], for resource-constrained mobile robots, recently, phenomenological and statistical modeling techniques such as kriging, Gaussian process regression, and kernel regression have gained much attention. Among phenomenological spatial models, adaptive control of multiple robotic sensors based on a parametric approach needs a persistent excitation (PE) condition for convergence of parameters [42, 50], while control strategies based on Bayesian spatial models do not require such conditions (e.g., by utilizing priori distributions as in Kalman filtering [41] or Gaussian process regression [2]). Hence, control engineers have become more aware of the usefulness of nonparametric Bayesian approaches such as Gaussian processes (defined by mean and covariance functions) [52, 53] to statistically model physical phenomena for the navigation of mobile sensor networks, e.g., [2–4, 36, 44–47]. Other more data-driven approaches have also developed (without statistical structure used in Gaussian processes) such as using kernel regression [48] and in reproducing kernel Hilbert spaces [49]. However, without a statistical structure in a random field, such an approach (as in [48, 49]) usually requires a great number of observations than the one with a statistical structure for a decent prediction quality.

In a mobile sensor network, the resource-limited sensing agents are required to collaborate in order to achieve a specific objective. The cooperative control becomes essential. The most popular applications are in networks of autonomous ground vehicles [54, 55], underwater and surface vehicles [23, 36, 56–58], or aerial vehicles [59–61]. Emerging technologies have been reported on the coordination of mobile sensing agents [41, 62–70].

The mobility of mobile agents can be designed in order to perform the optimal sampling of the field of interest. Optimal sampling design is the process of choosing where to take samples in order to maximize the information gained. Recently, in [36], Leonard et al. developed mobile sensor networks that optimize ocean sampling performance defined in terms of uncertainty in a model estimate of a sampled field. A typical sensor placement technique [71] that puts sensors at the locations where the entropy is high tends to place sensors along the borders of the area of interest [44]. In [44], Krause et al. showed that seeking sensor placements that are most informative about unsensed locations is NP-hard, and they presented a polynomial time approximation algorithm by exploiting the submodularity of mutual information [72]. In a similar approach, in [73], Singh et al. presented an efficient planning of informative paths for multiple robots that maximize the mutual information.

To find these locations that predict the phenomenon best, one needs a model of the spatiotemporal phenomenon. To this end, we use Gaussian processes (and Gaussian random fields) to model fields undergoing transport phenomena. Nonparametric Gaussian process regression (or Kriging in geostatistics) has been widely used as a nonlinear regression technique to estimate and predict geostatistical data [52, 53, 74, 75]. A Gaussian process is a natural generalization of the Gaussian probability distribution. It generalizes a Gaussian distribution with a finite number of random variables to a Gaussian process with an infinite number of random variables in the surveillance region [53]. Gaussian process modeling enables us to efficiently predict physical values, such as temperature, salinity, pH, or biomass of harmful algal blooms, at any point with a predicted uncertainty level. For instance, near-optimal static sensor placements with a mutual information criterion in Gaussian processes were proposed in [44, 76]. A distributed Kriged Kalman filter for spatial estimation based on mobile sensor networks was developed in [45]. Multiagent systems that are versatile for various tasks by exploiting predictive posterior statistics of Gaussian processes were developed in [77, 78].

Gaussian process regression, based on the standard mean and covariance functions, requires an inversion of a covariance matrix whose size grows as the number of observations increases. The significant computational complexity in Gaussian process regression due to the growing number of observations (and hence the size of covariance matrix) has been tackled in different ways [2, 79–83].

Unknown hyperparameters in the covariance function can be estimated by a maximum likelihood (ML) estimator or a maximum *a posteriori* (MAP) estimator and then be used in the prediction as the true hyperparameters [1]. However, the point estimate (ML or MAP estimate) itself needs to be identified using a sufficient amount of measurements and it fails to incorporate the uncertainty in the estimated hyperparameters into the prediction in a Bayesian perspective. The advantage of a fully Bayesian approach is that the uncertainty in the model parameters is incorporated in the prediction [84]. In [85], Gaudard et al. presented a Bayesian method that uses importance sampling for analyzing spatial data sampled from a Gaussian random field whose covariance function was unknown. However, the solution often requires Markov Chain Monte Carlo (MCMC) methods, which greatly increases the computational complexity. In [46], an iterative prediction algorithm without resorting to MCMC methods has been developed based on analytical closed-form solutions from results in [85], by assuming that the covariance function of the spatiotemporal Gaussian random field is known up to a constant.

There have been growing efforts to fit a computationally efficient Gaussian Markov random field (GMRF) on a discrete lattice to a Gaussian random field on a continuum space [4, 86–88]. It has been demonstrated that GMRFs with small neighborhoods can approximate Gaussian fields surprisingly well [86]. This approximated GMRF and its regression are very attractive for the resource-constrained mobile sensor networks due to its computational efficiency and scalability [89] as compared to the standard Gaussian process and its regression. Fast kriging of large datasets using a GMRF as an approximation of a Gaussian field has been proposed in [88].

1.2 Contents in Chapters

A brief summary for each subsequent chapter is as follows. Chapter 2 gives an introduction to Gaussian processes and Gaussian Markov random fields for general domains as well as the space-time domain.

In Chap. 3, we develop covariance function learning algorithms for the sensing agents to perform nonparametric prediction based on a properly adapted Gaussian process for a given spatiotemporal phenomenon. By introducing a generalized covariance function, we expand the class of Gaussian processes to include the anisotropic spatiotemporal phenomena. Maximum likelihood (ML) optimization is used to estimate hyperparameters for the associated covariance function as an empirical Bayes method. The proposed optimal navigation strategy for autonomous vehicles will maximize the Fisher information [90], improving the quality of the estimated covariance function.

In Chap. 4, we first present a theoretical foundation of Gaussian process regression with truncated observations. In particular, we show that the quality of prediction based on truncated observations does not deteriorate much as compared to that of prediction based on all cumulative data under certain conditions. The error bounds to use truncated observations are analyzed for prediction at a single point of interest. A way to select the temporal truncation size for spatiotemporal Gaussian processes is also introduced. Inspired by the analysis, we then propose both centralized and distributed navigation strategies for mobile sensor networks to move in order to reduce prediction error variances at points of interest. In particular, we demonstrate that the distributed navigation strategy produces an emergent, swarming-like, collective behavior to maintain communication connectivity among mobile sensing agents.

In Chap. 5, we formulate a fully Bayesian approach for spatiotemporal Gaussian process regression under practical conditions such as measurement noise and unknown hyperparameters (particularly, the bandwidths). Thus, multifactorial effects of observations, measurement noise, and prior distributions of hyperparameters are all correctly incorporated in the computed posterior predictive distribution. Using discrete prior probabilities and compactly supported kernels, we provide a way to design sequential Bayesian prediction algorithms that can be computed (without using the Gibbs sampler) in constant time (i.e., $O(1)$) as the number of observations increases. An adaptive sampling strategy for mobile sensors, using the maximum *a posteriori* (MAP) estimation, has been proposed to minimize the prediction error variances.

In Chap. 6, we propose a new class of Gaussian processes for resource-constrained mobile sensor networks that build on a Gaussian Markov random field (GMRF) with respect to a proximity graph over the surveillance region. The main advantages of using this class of Gaussian processes over standard Gaussian processes defined by mean and covariance functions are its numerical efficiency and scalability due to its built-in GMRF and its capability of representing a wide range of nonstationary physical processes. The formulas for predictive statistics are derived and a sequential field prediction algorithm is provided for sequentially sampled observations. For a special case using compactly supported weighting functions, we propose a distributed algorithm to implement field prediction by correctly fusing all observations.

In Chap. 7, we consider a discretized spatial field that is modeled by a GMRF with unknown hyperparameters. From a Bayesian perspective, we design a sequential prediction algorithm to exactly compute the predictive inference of the random field. The main advantages of the proposed algorithm are (1) the computational efficiency due to the sparse structure of the precision matrix, and (2) the scalability as the number of measurements increases. Thus, the prediction algorithm correctly takes into account the uncertainty in hyperparameters in a Bayesian way and also is scalable to be usable for the mobile sensor networks with limited resources. An adaptive sampling strategy is also designed for mobile sensing agents to find the most informative locations in taking future measurements in order to minimize the prediction error and the uncertainty in the estimated hyperparameters simultaneously.

Chapter 2
Preliminaries

2.1 Mathematical Notation

Standard notation is used throughout this book. Let $\mathbb{R}, \mathbb{R}_{\geq 0}, \mathbb{R}_{>0}, \mathbb{Z}, \mathbb{Z}_{\geq 0}, \mathbb{Z}_{>0}$ denote the sets of real numbers, nonnegative real numbers, positive real numbers, integers, nonnegative integers, and positive integers, respectively.

Let $\mathbb{E}$, Var, Corr, Cov denote the expectation, variance, correlation, and the covariance operators, respectively.

Let $\mathbf{A}^T \in \mathbb{R}^{M \times N}$ be the transpose of a matrix $\mathbf{A} \in \mathbb{R}^{N \times M}$. Let $\text{tr}(\mathbf{A})$ and $\det(\mathbf{A})$ denote the trace and the determinant of a matrix $\mathbf{A} \in \mathbb{R}^{N \times N}$, respectively. Let $\text{row}_i(\mathbf{A}) \in \mathbb{R}^M$ and $\text{col}_j(\mathbf{A}) \in \mathbb{R}^N$ denote the ith row and the jth column of a matrix $\mathbf{A} \in \mathbb{R}^{N \times M}$, respectively.

The positive definiteness and the positive semi-definiteness of a square matrix $\mathbf{A}$ are denoted by $\mathbf{A} \succ 0$ and $\mathbf{A} \succeq 0$, respectively.

Let $|x|$ denote the absolute value of a scalar x. Let $\|\mathbf{x}\|$ denote the standard Euclidean norm (2-norm) of a vector $\mathbf{x}$. The induced 2-norm of a matrix $\mathbf{A}$ is denoted by $\|\mathbf{A}\|$. Let $\|\mathbf{x}\|_\infty$ denote the infinity norm of a vector $\mathbf{x}$.

Let $\mathbf{1}$ denote the vector with all elements equal to one and $\mathbf{I}$ denote the identity matrix with an appropriate size. Let $\mathbf{e}_i$ be the standard basis vector of appropriate size with 1 as its ith element and 0 on all other elements.

The symbol $\otimes$ denotes the Kronecker product. The symbol $\circ$ denotes the Hadamard product (also known as the entry-wise product and the Schur product).

A random vector $\mathbf{x}$, which is distributed by a normal distribution of mean $\boldsymbol{\mu}$ and covariance matrix $\mathbf{C}$, is denoted by $\mathbf{x} \sim \mathbb{N}(\boldsymbol{\mu}, \mathbf{C})$. The corresponding probability density function is denoted by $\mathbb{N}(\mathbf{x}; \boldsymbol{\mu}, \mathbf{C})$.

The relative complement of a set $\mathcal{A}$ in a set $\mathcal{B}$ is denoted by $\mathcal{B} \setminus \mathcal{A} := \mathcal{B} \cap \mathcal{A}^c$, where $\mathcal{A}^c$ is the complement of $\mathcal{A}$. For a set $\mathcal{A} \in \mathcal{I}$, we define $z_\mathcal{A} = \{z_i \mid i \in \mathcal{A}\}$. Let $-\mathcal{A}$ denote the set $\mathcal{I} \setminus \mathcal{A}$.

Y. Xu et al., *Bayesian Prediction and Adaptive Sampling Algorithms for Mobile Sensor Networks*, SpringerBriefs in Control, Automation and Robotics, DOI 10.1007/978-3-319-21921-9_2

An undirected graph $\mathcal{G} = (\mathcal{V}, \mathcal{E})$ is a tuple consisting of a set of vertices $\mathcal{V} := \{1, \cdots, n\}$ and a set of edges $\mathcal{E} \subset \mathcal{V} \times \mathcal{V}$. The neighbors of $i \in \mathcal{V}$ in $\mathcal{G}$ are denoted by $\mathcal{N}_i := \{j \in \mathcal{V} \mid \{i, j\} \in \mathcal{E}\}$.

Other notations will be explained in due course.

2.2 Physical Process Model

In this section, we review important notions for the Gaussian process which will be used to model the physical phenomenon. In particular, we introduce a class of spatiotemporal Gaussian process model with anisotropic covariance functions. The properties of Gaussian Markov random fields (GMRF) are also briefly reviewed.

2.2.1 Gaussian Process

A Gaussian process can be thought of a generalization of a Gaussian distribution over a finite vector space to function space of infinite dimension. It is formally defined as follows [53, 91]:

Definition 2.1 A Gaussian process (GP) is a collection of random variables, any finite number of which have a consistent[1] joint Gaussian distribution.

A Gaussian process, denoted by

$$z(\mathbf{x}) \sim \mathcal{GP}\left(\mu(\mathbf{x}), C(\mathbf{x}, \mathbf{x}'; \boldsymbol{\theta})\right) \tag{2.1}$$

is completely specified by its mean function $\mu(\mathbf{x})$ and covariance function $C(\mathbf{x}, \mathbf{x}'; \boldsymbol{\theta})$ which are defined as

$$\begin{aligned} \mu(\mathbf{x}) &= \mathbb{E}\left[z(\mathbf{x})\right], \\ C(\mathbf{x}, \mathbf{x}'; \boldsymbol{\theta}) &= \mathbb{E}\left[(z(\mathbf{x}) - \mu(\mathbf{x}))\left(z(\mathbf{x}') - \mu(\mathbf{x}')\right)|\boldsymbol{\theta}\right]. \end{aligned}$$

Although not needed to be done, we take the mean function to be zero for notational simplicity,[2] i.e., $\mu(\mathbf{x}) = \mathbf{0}$. If the covariance function $C(\mathbf{x}, \mathbf{x}'; \boldsymbol{\theta})$ is invariant with respect to translations in the input space, i.e., $C(\mathbf{x}, \mathbf{x}'; \boldsymbol{\theta}) = C(\mathbf{x} - \mathbf{x}'; \boldsymbol{\theta})$, we call it stationary. Furthermore, if the covariance function is a function of only the distance between the inputs, i.e., $C(\mathbf{x}, \mathbf{x}'; \boldsymbol{\theta}) = C(\left\|\mathbf{x} - \mathbf{x}'\right\|; \boldsymbol{\theta})$, then it is called isotropic.

[1] It is also known as the marginalization property. It means simply that the random variables obey the usual rules of marginalization, etc.

[2] This is not a drastic limitation since the mean of the posterior process is not confined to zero [53].

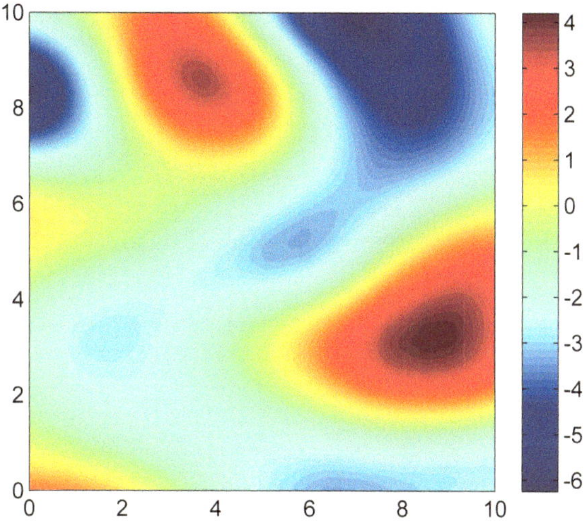

Fig. 2.1 Realization of a two-dimensional ($D = 2$) Gaussian process with $\sigma_f^2 = 5, \sigma_1 = 2.5$, and $\sigma_2 = 1.5$.

In practice, a parametric family of functions is used instead of fixing the covariance function [84]. One common choice of a stationary covariance function is

$$C(\mathbf{x}, \mathbf{x}'; \boldsymbol{\theta}) = \sigma_f^2 \exp\left\{ -\sum_{\ell=1}^{D} \frac{(x_\ell - x'_\ell)^2}{2\sigma_\ell^2} \right\}, \tag{2.2}$$

where x_ℓ is the ℓth element of $\mathbf{x} \in \mathbb{R}^D$. From (2.2), it can be easily seen that the correlation between two inputs decreases as the distance between them increases. This decreasing rate depends on the choice of the length scales $\{\sigma_\ell\}$. A very large length scale means that the predictions would have little bearing on the corresponding input which is then said to be insignificant. σ_f^2 gives the overall vertical scale relative to the mean of the Gaussian process in the output space. These parameters play the role of hyperparameters since they correspond to the hyperparameters in neural networks and in the standard parametric model. Therefore, we define $\boldsymbol{\theta} = (\sigma_f^2, \sigma_1, \cdots, \sigma_D)^T \in \mathbb{R}^{D+1}$ as the hyperparameter vector. A realization of a Gaussian process that is numerically generated is shown in Fig. 2.1.

2.2.2 Spatiotemporal Gaussian Process

In this section, spatiotemporal Gaussian processes are of particular interest. Spatiotemporal Gaussian processes are obtained as a special case of (2.1) by setting $\mathbf{x} \subset \mathbb{R}^D \times \mathbb{R}_{\geq 0}$, where $\mathbb{R}^D$ is for spatial locations and $\mathbb{R}_{\geq 0}$ is the temporal domain. A spatiotemporal Gaussian process can be written as

$$z(\mathbf{s}, t) \sim \mathcal{GP}(\mu(\mathbf{s}, t), C(\mathbf{s}, t, \mathbf{s}', t'; \boldsymbol{\theta})),$$

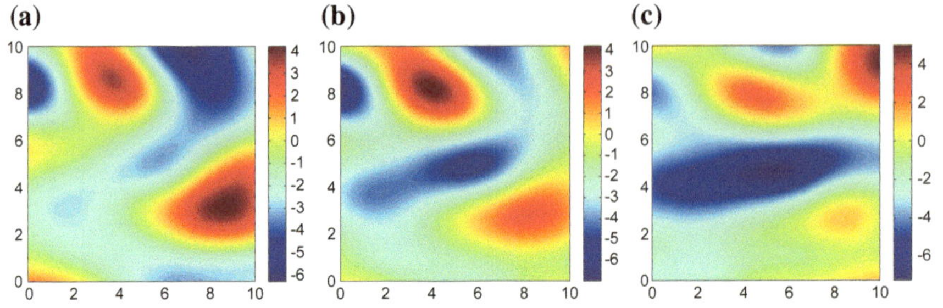

Fig. 2.2 Realization of a spatiotemporal ($D = 2$) Gaussian process with $\sigma_f^2 = 5$, $\sigma_1 = 2.5$, $\sigma_2 = 1.5$, and $\sigma_t = 8$ at **a** $t = 1$, **b** $t = 5$, and **c** $t = 10$.

where $\mathbf{x} = (\mathbf{s}^T, t)^T \in \mathbb{R}^D \times \mathbb{R}_{\geq 0}$. We consider the following generalized anisotropic covariance function $C(\mathbf{x}, \mathbf{x}'; \boldsymbol{\theta})$ with a hyperparameter vector $\boldsymbol{\theta} := (\sigma_f^2, \sigma_1, \cdots, \sigma_D, \sigma_t)^T \in \mathbb{R}^{D+2}$:

$$C(\mathbf{x}, \mathbf{x}'; \boldsymbol{\theta}) = \sigma_f^2 \exp\left(-\sum_{\ell=1}^{D} \frac{(s_\ell - s'_\ell)^2}{2\sigma_\ell^2}\right) \exp\left(-\frac{(t - t')^2}{2\sigma_t^2}\right), \tag{2.3}$$

where $\mathbf{s}, \mathbf{s}' \in \mathcal{Q} \subset \mathbb{R}^D$, $t, t' \in \mathbb{R}_{\geq 0}$. $\{\sigma_1, \cdots, \sigma_D\}$ and σ_t are kernel bandwidths for space and time, respectively. (2.3) shows that points close in the measurement space and time indices are strongly correlated and produce similar values. In reality, the larger temporal distance two measurements are taken with, the less correlated they become, which strongly supports our generalized covariance function in (2.3). This may also justify the truncation (or windowing) of the observed time series data to limit the size of the covariance matrix for reducing the computational cost. A spatially isotropic version of the covariance function in (2.3) has been used in [36]. A realization of a spatiotemporal Gaussian process that is numerically generated is shown in Fig. 2.2.

2.2.3 *Gaussian Markov Random Field*

The Gaussian Markov random field is formally defined as follows [92]:

Definition 2.2 *(GMRF, [92, Definition 2.1])* A random vector $\mathbf{z} = (z_1, \cdots, z_N)^T \in \mathbb{R}^N$ is called a GMRF with respect to a graph $\mathcal{G} = (\mathcal{V}, \mathcal{E})$ with mean $\boldsymbol{\mu}$ and precision matrix $\mathbf{Q} \succ 0$, if and only if its density has the form

$$\pi(\mathbf{z}) = \frac{|\mathbf{Q}|^{1/2}}{(2\pi)^{N/2}} \exp\left(-\frac{1}{2}(\mathbf{z} - \boldsymbol{\mu})^T \mathbf{Q}(\mathbf{z} - \boldsymbol{\mu})\right),$$

and $(\mathbf{Q})_{ij} \neq 0 \Leftrightarrow \{i, j\} \in \mathcal{E}$ for all $i \neq j$, where the precision matrix (or information matrix) $\mathbf{Q} = \mathbf{C}^{-1}$ is the inverse of the covariance matrix $\mathbf{C}$, and $|\mathbf{Q}|$ denotes the determinant of $\mathbf{Q}$.

The Markov property of a GMRF can be shown by the following theorem.

Theorem 2.1 ([92, Theorem 2.4]) *Let* $\mathbf{z}$ *be a GMRF with respect to* $\mathcal{G} = (\mathcal{V}, \mathcal{E})$. *Then the followings are equivalent.*

1. *The pairwise Markov property:*

$$z_i \perp z_j \mid z_{-ij} \quad \text{if } \{i, j\} \notin \mathcal{E} \text{ and } i \neq j,$$

where $\perp$ *denotes conditional independence and* $z_{-ij} := z_{-\{i,j\}} = z_{\mathcal{I} \setminus \{i,j\}}$. *This implies that* z_i *and* z_j *are conditionally independent given observations at all other vertices except* $\{i, j\}$ *if* i *and* j *are not neighbors.*

2. *The local Markov property:*

$$z_i \perp z_{-\{i, \mathcal{N}_i\}} \mid z_{\mathcal{N}_i} \quad \text{for every } i \in \mathcal{I}.$$

3. *The global Markov property:*

$$z_{\mathcal{A}} \perp z_{\mathcal{B}} \mid z_{\mathcal{C}}$$

for disjoint sets $\mathcal{A}$, $\mathcal{B}$, *and* $\mathcal{C}$ *where* $\mathcal{C}$ *separates* $\mathcal{A}$ *and* $\mathcal{B}$, *and* $\mathcal{A}$ *and* $\mathcal{B}$ *are nonempty.*

If a graph $\mathcal{G}$ has small cardinalities of the neighbor sets, its precision matrix $\mathbf{Q}$ becomes sparse with many zeros in its entries. This plays a key role in computation efficiency of a GMRF which can be greatly exploited by the resource-constrained mobile sensor network. For instance, some of the statistical inference can be obtained directly from the precision matrix $\mathbf{Q}$ with conditional interpretations.

Theorem 2.2 ([92, Theorem 2.3]) *Let* $\mathbf{z}$ *be a GMRF with respect to* $\mathcal{G} = (\mathcal{V}, \mathcal{E})$ *with mean* $\boldsymbol{\mu}$ *and precision matrix* $\mathbf{Q} \succ 0$, *then we have*

$$\begin{aligned}
\mathbb{E}(z_i \mid z_{-i}) &= \mu_i - \frac{1}{(\mathbf{Q})_{ii}} \sum_{j \in \mathcal{N}_i} (\mathbf{Q})_{ij}(z_j - \mu_j), \\
\mathrm{Var}(z_i \mid z_{-i}) &= \frac{1}{(\mathbf{Q})_{ii}}, \\
\mathrm{Corr}(z_i, z_j \mid z_{-ij}) &= -\frac{(\mathbf{Q})_{ij}}{\sqrt{(\mathbf{Q})_{ii}(\mathbf{Q})_{jj}}}, \quad \forall i \neq j.
\end{aligned}$$

2.3 Mobile Sensor Network

In this section, we explain the sensor network formed by multiple mobile sensing agents and present the measurement model used throughout the thesis.

Let N be the number of sensing agents distributed over the surveillance region $\mathcal{Q} \in \mathbb{R}^D$. The identity of each agent is indexed by $\mathcal{I} := \{1, 2, \cdots, N\}$. Assume that all agents are equipped with identical sensors and take noisy observations at time $t \in \mathbb{Z}_{>0}$. At time t, the sensing agent i takes a noise-corrupted measurement $y_i(t)$ at its current location $\mathbf{q}_i(t) \in \mathcal{Q}$, i.e.,

$$y_i(t) = z(\mathbf{q}_i(t), t) + \epsilon_i, \quad \epsilon_i \overset{i.i.d.}{\sim} \mathbb{N}(0, \sigma_w^2),$$

where the sensor noise ϵ_i is considered to be an independent and identically distributed Gaussian random variable. $\sigma_w^2 > 0$ is the noise level and we define the signal-to-noise ratio as

$$\gamma = \frac{\sigma_f^2}{\sigma_w^2}.$$

Notice that when a static field is considered, we have $z(\mathbf{s}, t) = z(\mathbf{s})$.

For notational simplicity, we denote the collection of positions of all N agents at time t as $\mathbf{q}(t)$, i.e.,

$$\mathbf{q}(t) := \left(\mathbf{q}_1(t)^T, \cdots, \mathbf{q}_N(t)^T\right)^T \in \mathcal{Q}^N.$$

The collective measurements from all N mobile sensors at time t are denoted by

$$\mathbf{y}_t := (y_1(t), \cdots, y_N(t))^T \in \mathbb{R}^N.$$

The cumulative measurements from time $t \in \mathbb{Z}_{>0}$ to time $t' \in \mathbb{Z}_{>0}$ are denoted by

$$\mathbf{y}_{t:t'} := \left(\mathbf{y}_t^T, \cdots, \mathbf{y}_{t'}^T\right)^T \in \mathbb{R}^{N(t'-t+1)}.$$

The communication network of mobile agents can be represented by an undirected graph. Let $\mathcal{G}(t) := (\mathcal{I}, \mathcal{E}(t))$ be an undirected communication graph such that an edge $(i, j) \in \mathcal{E}(t)$ if and only if agent i can communicate with agent $j \neq i$ at time t. We define the neighborhood of agent i at time t by $\mathcal{N}_i(t) := \{j \in \mathcal{I} \mid (i, j) \in \mathcal{E}(t)\}$. Similarly, let $\mathbf{q}^{[i]}(t)$ denote the vector form of the collection of positions in $\left\{\mathbf{q}_j(t) \mid j \in \{i\} \cup \mathcal{N}_i(t)\right\}$. Let $\mathbf{y}_t^{[i]}$ denote vector form of the collection of observations in $\left\{y(\mathbf{q}_j(t), t) \mid j \in \{i\} \cup \mathcal{N}_i(t)\right\}$. The cumulative measurements of agent i from time t to time t' are denoted as $\mathbf{y}_{t:t'}^{[i]}$.

2.4 Gaussian Processes for Regression

Suppose we have a dataset $\mathcal{D} = \left\{(\mathbf{x}^{(i)}, y^{(i)}) \mid i = 1, \cdots, n\right\}$ collected by mobile sensing agents where $\mathbf{x}^{(i)}$ denotes an input vector of dimension D and $y^{(i)}$ denotes a scalar value of the noise-corrupted output. The objective of probabilistic regression is to compute the predictive distribution of the function values $z_* := z(\mathbf{x}_*)$ at some test input $\mathbf{x}_*$.

For notational simplicity, we define the design matrix $\mathbf{X}$ of dimension $n \times D$ as the aggregation of n input vectors (i.e., $\text{row}_i(\mathbf{X}) := (\mathbf{x}^{(i)})^T$), and the outputs are collected in a vector $\mathbf{y} := (y^{(1)}, \cdots, y^{(n)})^T$. The corresponding vector of noise-free outputs is defined as $\mathbf{z} := (z(\mathbf{x}^{(1)}), \cdots, z(\mathbf{x}^{(n)}))^T$.

The advantage of the Gaussian process formulation is that the combination of the prior and noise models can be carried out exactly via matrix operations [93]. The idea of Gaussian process regression is to place a GP prior directly on the space of functions without parameterizing the function $z(\cdot)$, i.e.,

$$\pi(\mathbf{z}|\boldsymbol{\theta}) = \mathbb{N}(\mathbf{z}; \boldsymbol{\mu}, \mathbf{K}),$$

where $\boldsymbol{\mu} \in \mathbb{R}^n$ is the mean vector obtained by $(\boldsymbol{\mu})_i = \mu(\mathbf{x}^{(i)})$, and $\mathbf{K} := \text{Cov}(\mathbf{z}, \mathbf{z}|\boldsymbol{\theta}) \in \mathbb{R}^{n \times n}$ is the covariance matrix obtained by $(\mathbf{K})_{ij} = C(\mathbf{x}^{(i)}, \mathbf{x}^{(j)}; \boldsymbol{\theta})$. Notice that the GP model and all expressions are always conditional on the corresponding inputs. In the following, we will always neglect the explicit conditioning on the input matrix $\mathbf{X}$.

The inference in the Gaussian process model is as follows. First, we assume a joint GP prior $\pi(\mathbf{z}, z_*|\boldsymbol{\theta})$ over functions, i.e.,

$$\pi(\mathbf{z}, z_*|\boldsymbol{\theta}) = \mathbb{N}\left(\begin{bmatrix} \boldsymbol{\mu} \\ \mu(\mathbf{x}_*) \end{bmatrix}, \begin{bmatrix} \mathbf{K} & \mathbf{k} \\ \mathbf{k}^T & C(\mathbf{x}_*, \mathbf{x}_*; \boldsymbol{\theta}) \end{bmatrix}\right), \tag{2.4}$$

where $\mathbf{k} := \text{Cov}(\mathbf{z}, z_*|\boldsymbol{\theta}) \in \mathbb{R}^n$ is the covariance between $\mathbf{z}$ and z_* obtained by $(\mathbf{k})_i = C(\mathbf{x}^{(i)}, \mathbf{x}_*; \boldsymbol{\theta})$. Then, the joint posterior is obtained using Bayes rule, i.e.,

$$\pi(\mathbf{z}, z_*|\boldsymbol{\theta}, \mathbf{y}) = \frac{\pi(\mathbf{y}|\mathbf{z})\pi(\mathbf{z}, z_*|\boldsymbol{\theta})}{\pi(\mathbf{y}|\boldsymbol{\theta})},$$

where we have used $\pi(\mathbf{y}|\mathbf{z}, z_*) = \pi(\mathbf{y}|\mathbf{z})$. Finally, the desired predictive distribution $\pi(z_*|\boldsymbol{\theta}, \mathbf{y})$ is obtained by marginalizing out the latent variables in $\mathbf{z}$, i.e.,

$$\begin{aligned} \pi(z_*|\boldsymbol{\theta}, \mathbf{y}) &= \int \pi(\mathbf{z}, z_*|\theta, \mathbf{y}) d\mathbf{z} \\ &= \frac{1}{\pi(\mathbf{y}|\boldsymbol{\theta})} \int \pi(\mathbf{y}|\mathbf{z})\pi(\mathbf{z}, z_*|\boldsymbol{\theta}, \mathbf{y}) d\mathbf{z}. \end{aligned} \tag{2.5}$$

Since we have the joint Gaussian prior given in (2.4) and

$$\mathbf{y}|\mathbf{z} \sim \mathbb{N}\left(\mathbf{z}, \sigma_w^2 \mathbf{I}\right),$$

the integral in (2.5) can be evaluated in closed-form and the predictive distribution turns out to be Gaussian, i.e.,

$$z_*|\boldsymbol{\theta}, \mathbf{y} \sim \mathbb{N}\left(\mu_{z_*|\boldsymbol{\theta},\mathbf{y}}, \sigma^2_{z_*|\boldsymbol{\theta},\mathbf{y}}\right), \tag{2.6}$$

where

$$\mu_{z_*|\boldsymbol{\theta},\mathbf{y}} = \mu(\mathbf{x}_*) + \mathbf{k}^T(\mathbf{K} + \sigma_w^2 \mathbf{I})^{-1}(\mathbf{y} - \boldsymbol{\mu}), \tag{2.7}$$

and

$$\sigma^2_{z_*|\boldsymbol{\theta},\mathbf{y}} = C(\mathbf{x}_*, \mathbf{x}_*; \boldsymbol{\theta}) - \mathbf{k}^T(\mathbf{K} + \sigma_w^2 \mathbf{I})^{-1}\mathbf{k}. \tag{2.8}$$

For notational simplicity, we define the covariance matrix of the noisy observations as $\mathbf{C} := \mathrm{Cov}(\mathbf{y}, \mathbf{y}|\boldsymbol{\theta}) = \mathbf{K} + \sigma_w^2 \mathbf{I}$.

Chapter 3
Learning Covariance Functions

We often assume that Gaussian processes are isotropic implying that the covariance function only depends on the distance between locations. Many studies also assume that the corresponding covariance functions are known *a priori* for simplicity. However, this is not the case in general as pointed out in the literature [44, 76, 94], in which they treat the nonstationary process by fusing a collection of isotropic spatial Gaussian processes associated with a set of local regions. Our objective in this chapter is to develop theoretically sound algorithms for mobile sensor networks to learn the anisotropic covariance function of a spatiotemporal Gaussian process. Mobile sensing agents can then predict the Gaussian process based on the estimated covariance function in a nonparametric manner.

First, in Sect. 3.1, we illustrate the importance of the choice of the covariance function. In Sect. 3.2, we introduce a covariance function learning algorithm for an anisotropic, spatiotemporal Gaussian process. The covariance function is assumed to be deterministic but unknown *a priori* and it is estimated by the maximum likelihood (ML) estimator. In Sect. 3.3, an optimal sampling strategy is proposed to minimize the Cramér–Rao lower bound (CRLB) of the estimation error covariance matrix. In Sect. 3.4, simulation results illustrate the usefulness of our proposed approach.

3.1 Selection of Gaussian Process Prior

In this subsection, we illustrate the importance of selecting a Gaussian process prior via hyperparameters when we make inferences from the experimental data. To have illustrative cases, we consider the experimental data collected by the robotic boat (see Fig. 3.1a) that was deployed in a pond in Central Park, Okemos, Michigan. A set of water depth values and sampling locations was collected from onboard sensors in the robotic boat. Figure 3.1b shows the location site with boat trajectories in red lines. Without loss of generality, the GPS data, in particular the longitude and latitude, are normalized to $[0,\ 1]$. Let us assume that the process has a known constant mean so

Y. Xu et al., *Bayesian Prediction and Adaptive Sampling Algorithms for Mobile Sensor Networks*, SpringerBriefs in Control, Automation and Robotics, DOI 10.1007/978-3-319-21921-9_3

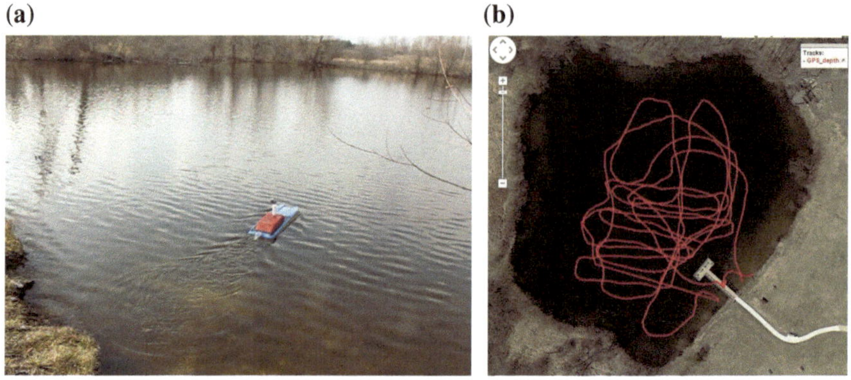

Fig. 3.1 **a** Remotely controlled boat equipped with depth sensor and GPS (credit: Jongeun Choi), **b** experiment site with robotic boat's trajectories (shown as *red lines*)

that we only care to select the covariance function. With the squared exponential covariance function in (2.2), i.e.,

$$C(\mathbf{x}, \mathbf{x}'; \theta) = \sigma_f^2 \exp\left\{ -\sum_{\ell=1}^{2} \frac{(x_\ell - x'_\ell)^2}{2\sigma_\ell^2} \right\},$$

the estimated depth field and the prediction error variance using the estimated hyperparameters by maximizing the likelihood function are shown in Fig. 3.2. Another set of corresponding figures are provided with the scaled σ_1 and σ_2 values by 0.2 from the estimates as shown in Fig. 3.3. With different bandwidths (or length scales) σ_1 and σ_2 while keeping other parameters fixed on the same values, the prediction and its prediction error variance on the same experimental data are significantly different.

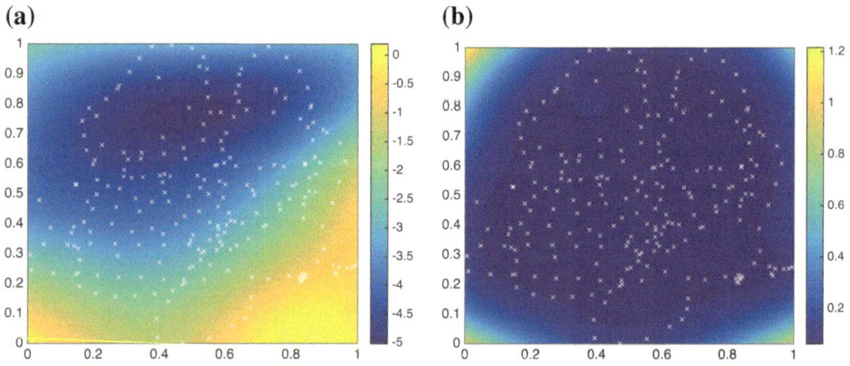

Fig. 3.2 Prediction with estimated hyperparameters $\sigma_f = 2.19, \sigma_1 = 0.40, \sigma_2 = 0.29, \sigma_w = 0.24$. **a** Estimated depth, and **b** prediction error variance, with sampling positions shown as white crosses

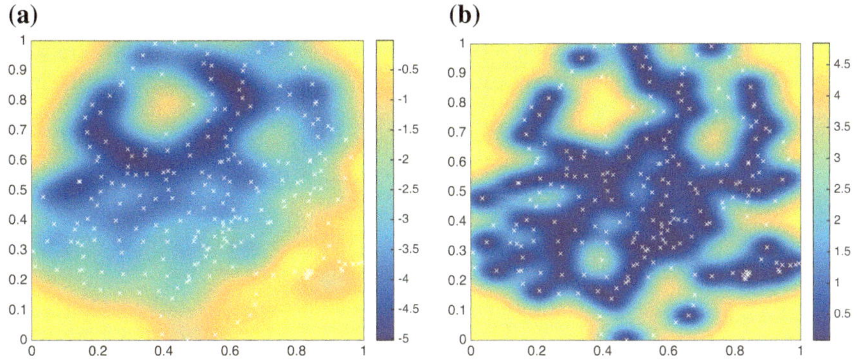

Fig. 3.3 Prediction with scaled σ_1 and σ_2 by 0.2, i.e., $\sigma_f = 2.19$, $\sigma_1 = 0.08$, $\sigma_2 = 0.058$, $\sigma_w = 0.24$. **a** Estimated depth, and **b** prediction error variance, with sampling positions shown as white crosses

With smaller bandwidths, the predicted depth field is much more wiggly and wavy than its counterpart (Figs. 3.2a and 3.3a), which is an artifact due to the wrong choice of bandwidths in this case.

For the sake of illustration, let us assume that hyperparmeters are known to be different as reported in Figs. 3.2 and 3.3. As shown in Figs. 3.2 and 3.3, in general, the Gaussian process with larger bandwidths (and stronger spatial correlations) tends to be smooth and does not need dense sampling for a decent level of prediction error variance quality (Fig. 3.2b). On the other hand, the Gaussian process with smaller bandwidths (and weaker spatial correlations) allows much more complicated spatial details and needs to be densely sampled. From these findings, we recognize the importance of the choice of the covariance function to make precise prediction as well as schedule the sampling in an optimal way. More discussions on the selection of covariance functions from a Bayesian perspective can be found in [95].

3.2 Learning the Hyperparameters

Without loss of generality, we consider a zero-mean spatiotemporal Gaussian process

$$z(\mathbf{s}, t) \sim \mathcal{GP}\left(0, C(\mathbf{s}, t, \mathbf{s}', t'; \theta)\right),$$

with the covariance function

$$C(\mathbf{s}, t, \mathbf{s}', t'; \theta) = \sigma_f^2 \exp\left(-\sum_{\ell=1,2} \frac{(s_\ell - s'_\ell)^2}{2\sigma_\ell^2}\right) \exp\left(-\frac{(t - t')^2}{2\sigma_t^2}\right),$$

where $\mathbf{s}, \mathbf{s}' \in \mathcal{Q} \subset \mathbb{R}^2$, $t, t' \in \mathbb{R}_{\geq 0}$, for modeling the field undergoing a physical transport phenomenon. $\boldsymbol{\theta} = (\sigma_f^2, \sigma_1, \sigma_2, \sigma_t)^T \in \mathbb{R}^M$ is the hyperparameter vector, where $M = 4$. The assumption of zero-mean is not a strong limitation since the mean of the posterior process is not confined to zero [53].

If the covariance function $C(\mathbf{s}, t, \mathbf{s}', t'; \boldsymbol{\theta})$ of a Gaussian process is not known *a priori*, mobile agents need to estimate parameters of the covariance function (i.e., the hyperparameter vector $\boldsymbol{\theta} \in \mathbb{R}^M$) based on the observed samples. In the case where measurement noise level σ_w is also unknown, it can be incorporated in the hyperparameter vector and be estimated. Thus, we have $\boldsymbol{\theta} = (\sigma_f^2, \sigma_1, \sigma_2, \sigma_t, \sigma_w)^T \in \mathbb{R}^M$ where $M = 5$.

Existing techniques for learning the hyperparameters are based on the likelihood function. Given the observations $\mathbf{y} = (y^{(1)}, \ldots, y^{(n)})^T \in \mathbb{R}^n$ collected by mobile sensing agents over time, the likelihood function is defined as

$$L(\boldsymbol{\theta}|\mathbf{y}) = \pi(\mathbf{y}|\boldsymbol{\theta}). \tag{3.1}$$

Notice that in this chapter, the hyperparameter vector $\boldsymbol{\theta}$ is considered to be deterministic, and hence $\pi(\mathbf{y}|\boldsymbol{\theta})$ should not be considered as conditional distribution.

A point estimate of the hyperparameter vector $\boldsymbol{\theta}$ can be made by maximizing the log likelihood function. The maximum likelihood (ML) estimate $\hat{\boldsymbol{\theta}} \in \mathbb{R}^M$ of the hyperparameter vector is obtained by

$$\hat{\boldsymbol{\theta}} = \arg\max_{\boldsymbol{\theta} \in \boldsymbol{\Theta}} \log L(\boldsymbol{\theta}|\mathbf{y}), \tag{3.2}$$

where $\boldsymbol{\Theta}$ is the set of all possible choices of $\boldsymbol{\theta}$. The log likelihood function is given by

$$\log L(\boldsymbol{\theta}|\mathbf{y}) = -\frac{1}{2}\mathbf{y}^T\mathbf{C}^{-1}\mathbf{y} - \frac{1}{2}\log\det(\mathbf{C}) - \frac{n}{2}\log 2\pi,$$

where $\mathbf{C} := \mathrm{Cov}(\mathbf{y}, \mathbf{y}|\boldsymbol{\theta}) \in \mathbb{R}^{n\times n}$ is the covariance matrix, and n is the total number of observations. Maximization of the log likelihood function can be done efficiently using gradient-based optimization techniques such as the conjugate gradient method [96, 97]. The partial derivative of the log likelihood function with respect to a hyperparameter $\theta_i \in \mathbb{R}$, i.e., the ith entry of the hyperparameter vector $\boldsymbol{\theta}$, is given by

$$\begin{aligned}\frac{\partial \log L(\boldsymbol{\theta}|\mathbf{y})}{\partial \theta_i} &= \frac{1}{2}\mathbf{y}^T\mathbf{C}^{-1}\frac{\partial \mathbf{C}}{\partial \theta_i}\mathbf{C}^{-1}\mathbf{y} - \frac{1}{2}\operatorname{tr}\left(\mathbf{C}^{-1}\frac{\partial \mathbf{C}}{\partial \theta_i}\right)\\ &= \frac{1}{2}\operatorname{tr}\left((\alpha\alpha^T - \mathbf{C}^{-1})\frac{\partial \mathbf{C}}{\partial \theta_i}\right),\end{aligned}$$

where $\alpha = \mathbf{C}^{-1}\mathbf{y} \in \mathbb{R}^n$. In general, the log likelihood function is a nonconvex function and hence it can have multiple maxima.

As an alternative, when certain prior knowledge is available on the hyperparameters, a prior distribution $\pi(\boldsymbol{\theta})$ can be imposed on the hyperparameter vector. Using

Bayes' rule, the posterior distribution $\pi(\boldsymbol{\theta}|\mathbf{y})$ is proportional to the likelihood $L(\boldsymbol{\theta}|\mathbf{y})$ times the prior distribution $\pi(\boldsymbol{\theta})$, i.e.,

$$\begin{aligned}\pi(\boldsymbol{\theta}|\mathbf{y}) &\propto L(\boldsymbol{\theta}|\mathbf{y})\pi(\boldsymbol{\theta}) \\ &= \pi(\mathbf{y}|\boldsymbol{\theta})\pi(\boldsymbol{\theta}),\end{aligned}$$

in light of (3.1) defined earlier. Then the maximum *a posteriori* (MAP) estimate $\hat{\boldsymbol{\theta}} \in \mathbb{R}^M$ of the hyperparameter vector can be obtained similarly by

$$\hat{\boldsymbol{\theta}} = \arg\max_{\boldsymbol{\theta}\in\boldsymbol{\Theta}} \left(\log L(\boldsymbol{\theta}|\mathbf{y}) + \log \pi(\boldsymbol{\theta})\right). \tag{3.3}$$

Notice that when no prior information is available, the MAP estimate is equivalent to the ML estimate.

Once the estimate of the hyperparameter vector $\boldsymbol{\theta}$ is obtained with confidence, it can be used as the true value for the mobile sensor network to predict the field of interest using Gaussian process regression in (2.6).

3.3 Optimal Sampling Strategy

We assume now that the estimate of hyperparameters has been obtained using the procedure outlined in Sect. 3.2 based on all observations collected up to and including time t. At time $t+1$, agents should find new sampling positions to improve the quality of the estimated covariance function. For instance, to precisely estimate the anisotropic phenomenon, i.e., processes with different covariances along x and y directions, sensing agents need to explore and sample measurements along different directions.

To this end, we consider a centralized scheme. Suppose that a central station (or a leader agent) has access to all measurements collected by agents. Assume that at time $t+1$, agent i moves to a new sampling position $\tilde{\mathbf{q}}_i \in \mathcal{Q}$ and makes an observation $y_i(t+1) \in \mathbb{R}$. The collection of the new sampling positions and new observations from all agents are denoted by $\tilde{\mathbf{q}} \in \mathcal{Q}^N$ and $\tilde{\mathbf{y}} \in \mathbb{R}^N$, respectively. The objective of the optimal sampling strategy is to find the best sampling positions $\tilde{\mathbf{q}}$ such that the maximum likelihood (ML) estimate $\hat{\boldsymbol{\theta}}_{t+1} \in \mathbb{R}^M$ at time $t+1$ is as close to the true hyperparameter vector $\boldsymbol{\theta}^* \in \mathbb{R}^M$ as possible. Notice that $\boldsymbol{\theta}^*$ is unknown in practice, but this issue will be addressed slightly later.

Consider the Fisher information matrix (FIM) that measures the information produced by $\mathbf{y}_{1:t} \in \mathbb{R}^{Nt}$ and $\tilde{\mathbf{y}} \in \mathbb{R}^N$ for estimating the true hyperparameter vector $\boldsymbol{\theta}^* \in \mathbb{R}^M$ at time $t+1$. The Cramér–Rao lower bound (CRLB) theorem states that the inverse of the FIM (denoted by $\mathbf{M} \in \mathbb{R}^{M\times M}$) is a lower bound of the estimation error covariance matrix [90, 98]:

$$\mathbb{E}\left[(\hat{\boldsymbol{\theta}}_{t+1} - \boldsymbol{\theta}^*)(\hat{\boldsymbol{\theta}}_{t+1} - \boldsymbol{\theta}^*)^T\right] \succeq \mathbf{M}^{-1},$$

where $\hat{\boldsymbol{\theta}}_{t+1} \in \mathbb{R}^m$ represents the ML estimate of $\boldsymbol{\theta}^*$ at time $t+1$. The FIM [90] is given by

$$(\mathbf{M})_{ij} = -\mathbb{E}\left[\frac{\partial^2 \ln L(\boldsymbol{\theta}|\tilde{\mathbf{y}}, \mathbf{y}_{1:t})}{\partial\theta_i \partial\theta_j}\right],$$

where $L(\boldsymbol{\theta}|\tilde{\mathbf{y}}, \mathbf{y}_{1:t})$ is the likelihood function at time $t+1$, and the expectation is taken with respect to $\pi(\mathbf{y}_{1:t}, \tilde{\mathbf{y}}|\boldsymbol{\theta})$. Notice that the likelihood is now a function of $\boldsymbol{\theta}$ and $\tilde{\mathbf{y}}$. The analytical form of the FIM is given by

$$(\mathbf{M})_{ij} = \frac{1}{2}\operatorname{tr}\left(\tilde{\mathbf{C}}^{-1}\frac{\partial\tilde{\mathbf{C}}}{\partial\theta_i}\tilde{\mathbf{C}}^{-1}\frac{\partial\tilde{\mathbf{C}}}{\partial\theta_j}\right),$$

where $\tilde{\mathbf{C}} \in \mathbb{R}^{N(t+1)\times N(t+1)}$ is defined as

$$\tilde{\mathbf{C}} := \operatorname{Cov}\left(\begin{bmatrix}\mathbf{y}_{1:t}\\ \tilde{\mathbf{y}}\end{bmatrix}, \begin{bmatrix}\mathbf{y}_{1:t}\\ \tilde{\mathbf{y}}\end{bmatrix}\middle|\boldsymbol{\theta}^*\right).$$

Since the true value $\boldsymbol{\theta}^*$ is not available, we will evaluate the FIM at the currently best estimate $\hat{\boldsymbol{\theta}}_t$.

We can expect that minimizing the Cramér–Rao lower bound results in a decrease of uncertainty in estimating $\boldsymbol{\theta}$ [99]. The most common optimality criterion is D-optimality [100, 101]. It corresponds to minimizing the volume of the ellipsoid which represents the maximum confidence region for the maximum likelihood estimate of the unknown hyperparameters [101]. Using the D-optimality criterion [100, 101], the objective function $J(\cdot)$ is given by

$$J(\tilde{\mathbf{q}}) := \det(\mathbf{M}^{-1}).$$

However, if one hyperparameter has a very large variance compared to the others, the ellipsoid will be skinny and thus minimizing the volume may be misleading [101]. As an alternative, A-optimality which minimizes the sum of the variances is often used. The objective function $J(\cdot)$ based on A-optimality criterion is

$$J(\tilde{\mathbf{q}}) := \operatorname{tr}(\mathbf{M}^{-1}).$$

Hence, a control law for the mobile sensor network can be formulated as follows:

$$\mathbf{q}(t+1) = \arg\min_{\tilde{\mathbf{q}}\in\mathcal{Q}^N} J(\tilde{\mathbf{q}}). \tag{3.4}$$

In (3.4), we only consider the constraint that robots should move within the region $\mathcal{Q}$. However, the mobility constraints, such as the maximum distance that a robot

Table 3.1 Centralized optimal sampling strategy at time t

For $i \in \mathcal{I}$, agent i performs:

1: make an observation at current position $\mathbf{q}_i(t)$, *i.e.*, $y_i(t)$
2: transmit the observation $y_i(t)$ to the central station

The central station performs:

1: collect the observations from all N agents, *i.e.*, $\mathbf{y}_t$
2: obtain the cumulative measurements, *i.e.*, $\mathbf{y}_{1:t}$
3: compute the maximum likelihood estimate $\hat{\boldsymbol{\theta}}_t$ based on

$$\hat{\boldsymbol{\theta}}_t = \arg\max_{\boldsymbol{\theta} \in \boldsymbol{\Theta}} \ln L(\boldsymbol{\theta} | \mathbf{y}_{1:t}),$$

starting with the initial point $\hat{\boldsymbol{\theta}}_{t-1}$
4: compute the control in order to minimize the cost function $J(\tilde{q})$ via

$$\mathbf{q}(t+1) = \arg\min_{\tilde{\mathbf{q}} \in \mathcal{Q}^N} J(\tilde{\mathbf{q}})$$

5: send the next sampling positions $\{\mathbf{q}_i(t+1) \,|\, i \in \mathcal{I}\}$ to all N agents

For $i \in \mathcal{I}$, agent i performs:

1: receive the next sampling position $\mathbf{q}_i(t+1)$ from the central station
2: move to $\mathbf{q}_i(t+1)$ before time $t+1$

can move between two time indices, or the maximum speed with which a robot can travel, can be incorporated as additional constraints in the optimization problem [45].

The overall protocol for the sensor network is summarized as in Table 3.1.

3.4 Simulation

We apply our approach to a spatial Gaussian process. The Gaussian process was numerically generated for the simulation [53]. The hyperparameters used in the simulation were chosen such that $\boldsymbol{\theta} = (\sigma_f^2, \sigma_1, \sigma_2, \sigma_w)^T = (5, 4, 2, 0.5)^T$. In this case, $N = 9$ mobile sensing agents were initialized at random positions in a surveillance region $\mathcal{Q} = [0, 10] \times [0, 10]$. The initial values for the algorithm were given to be $\theta_0 = (1, 1, 1, 0.1)^T$. The gradient method was used to find the MAP estimate of the hyperparameter vector.

For simplicity, we assumed that the global basis is the same as the model basis. We considered a situation where at each time, measurements of agents are transmitted to a leader (or a central station) that uses our Gaussian learning algorithm and sends optimal control back to individual agents for next iteration to improve the quality of the estimated covariance function. The maximum distance for agents to move in one time step was chosen to be 1 for both x and y directions. The A-optimality criterion was used for optimal sampling.

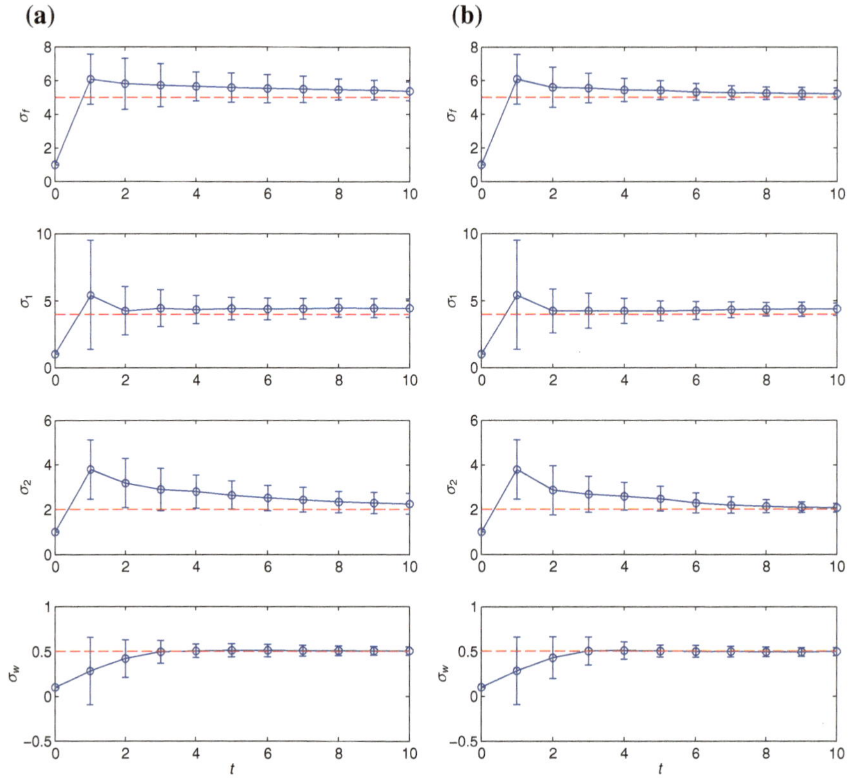

Fig. 3.4 Monte Carlo simulation results (100 runs) for a spatiotemporal Gaussian process using **a** the random sampling strategy, and **b** the adaptive sampling strategy. The estimated hyperparameters are shown in *blue circles* with error bars. The true hyperparameters that used for generating the process are shown in *red dashed lines*

For both proposed and random strategies, Monte Carlo simulations were run for 100 times and the statistical results are shown in Fig. 3.4. The estimates of the hyperparameters (shown in circles and error bars) tend to converge to the true values (shown in dotted lines) for both strategies. As can be seen, the proposed scheme (Fig. 3.4a) outperforms the random strategy (Fig. 3.4b) in terms of the A-optimality criterion.

After converging to a good estimate of $\boldsymbol{\theta}$, agents can switch to a decentralized configuration and collect samples for other goals such as peak tracking and prediction of the process [42, 77, 78].

Chapter 4
Memory Efficient Prediction With Truncated Observations

The main reason why the nonparametric prediction using Gaussian processes has not been popular for resource-constrained multi-agent systems is the fact that the optimal prediction must use all cumulatively measured values in a non-trivial way [74, 75]. In this case, a robot needs to compute the inverse of the covariance matrix whose size grows as it collects more measurements. With this operation, the robot will run out of memory quickly. Therefore, it is necessary to develop a class of prediction algorithms using spatio-temporal Gaussian processes under a fixed memory size.

A simple way to cope with this dilemma is to design a robot so that it predicts a spatio-temporal Gaussian process at the current (or future) time based on truncated observations, e.g., the last m observations from a total of n of observations as shown in Fig. 4.1. This seems intuitive in the sense that the last m observations are more correlated with the point of interest than the other $r = n - m$ observations (Fig. 4.1) in order to predict values at current or future time. Therefore, it is very important to analyze the performance degradation and trade-off effects of prediction based on truncated observations compared to the one based on all cumulative observations.

The second motivation is to design and analyze distributed sampling strategies for resource-constrained mobile sensor networks. Developing distributed estimation and coordination algorithms for multi-agent systems using only local information from local neighboring agents has been one of the most fundamental problems in mobile sensor networks [42, 45, 62–66]. Emphasizing practicality and usefulness, it is critical to synthesize and analyze distributed sampling strategies under practical constraints such as measurement noise and a limited communication range.

In Sect. 4.1, we propose to use only truncated observations to bound the computational complexity. The error bounds in using truncated observations are analyzed for prediction at a single point in Sect. 4.1.1. A way of selecting a temporal truncation size is also discussed in Sect. 4.1.2. To improve the prediction quality, centralized and distributed navigation strategies for mobile sensor networks are proposed in Sect. 4.2. In Sect. 4.3, simulation results illustrate the usefulness of our schemes under different conditions and parameters.

Y. Xu et al., *Bayesian Prediction and Adaptive Sampling Algorithms for Mobile Sensor Networks*, SpringerBriefs in Control, Automation and Robotics, DOI 10.1007/978-3-319-21921-9_4

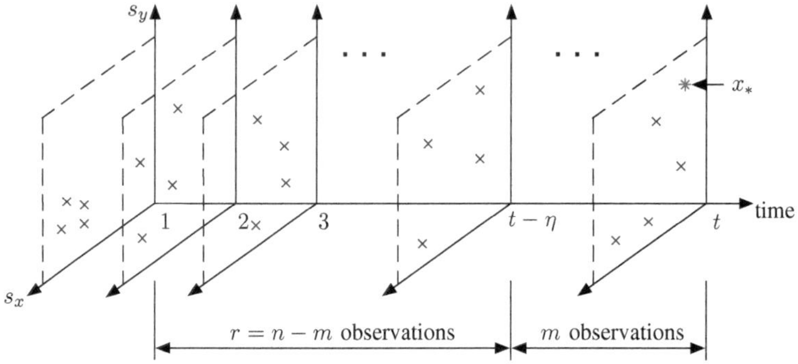

Fig. 4.1 Robot predicts a scalar value at $\mathbf{x}_*$ (denoted by a *red star*) based on cumulative n spatio-temporal observations (denoted by *blue crosses*). Near-optimal prediction can be obtained using truncated observations, e.g., the last m observations. In this case, $\mathbf{x} = (s_x, s_y, t)^T$

4.1 GPR with Truncated Observations

As mentioned in above, one drawback of Gaussian process regression is that its computational complexity and memory space increase as more measurements are collected, making the method prohibitive for robots with limited memory and computing power. To overcome this increase in complexity, a number of approximation methods for Gaussian process regression have been proposed. In particular, the sparse greedy approximation method [79], the Nystrom method [80], the informative vector machine [81], the likelihood approximation [82], and the Bayesian committee machine [83] have been shown to be effective for many problems. However, these approximation methods have been proposed without theoretical justifications.

In general, if measurements are taken from nearby locations (or space-time locations), correlation between measurements is strong and correlation exponentially decays as the distance between locations increases. If the correlation function of a Gaussian process has this property, intuitively, we can make a good prediction at a point of interest using only measurements nearby. In the next subsection, we formalize this idea and provide a theoretical foundation for justifying Gaussian process regression with truncated observations proposed in this chapter.

4.1.1 Error Bounds Using Truncated Observations

Consider a zero-mean Gaussian process

$$z(\mathbf{x}) \sim \mathcal{GP}(0, \sigma_f^2 C(\mathbf{x}, \mathbf{x}')). \tag{4.1}$$

Notice that we denote the covariance function as $\sigma_f^2 C(\mathbf{x}, \mathbf{x}')$ in which $C(\mathbf{x}, \mathbf{x}') := \mathrm{Corr}(z(\mathbf{x}), z(\mathbf{x}'))$ is the correlation function. Recall that the predictive distribution of $z_* := z(\mathbf{x}_*)$ at a point of interest $\mathbf{x}_*$ given observations $\mathbf{y} = (y^{(1)}, \ldots, y^{(n)})^T$ is Gaussian, i.e.,

$$z_*|\mathbf{y} \sim \mathbb{N}\left(\mu_{z_*|\mathbf{y}}, \sigma^2_{z_*|\mathbf{y}}\right), \tag{4.2}$$

where

$$\mu_{z_*|\mathbf{y}} = \mathbf{k}^T \mathbf{C}^{-1} \mathbf{k}, \tag{4.3a}$$

and

$$\sigma^2_{z_*|\mathbf{y}} = \sigma_f^2 (1 - \mathbf{k}^T \mathbf{C}^{-1} \mathbf{k}). \tag{4.3b}$$

In (4.3a) and (4.3b), we have defined $\mathbf{C} := \mathrm{Corr}(\mathbf{y}, \mathbf{y}) \in \mathbb{R}^{n \times n}$, and $\mathbf{k} := \mathrm{Corr}(\mathbf{y}, z_*) \in \mathbb{R}^n$. Notice that in this chapter, we assume the hyperparameter vector $\theta \in \mathbb{R}^M$ is given, and hence we neglect the explicit conditioning on θ.

Without loss of generality, we assume that the first m out of n observations are used to predict z_*. Let $r = n - m$, $\mathbf{y}_m = (y^{(1)}, \ldots, y^{(m)})^T$, $\mathbf{y}_r = (y^{(m+1)}, \ldots, y^{(n)})^T$. Then the covariance matrix $\mathbf{K} \in \mathbb{R}^{n \times n}$ and $\mathbf{k} \in \mathbb{R}^n$ can be represented as

$$\mathbf{K} = \begin{bmatrix} \mathbf{K}_m & \mathbf{K}_{mr} \\ \mathbf{K}_{mr}^T & \mathbf{K}_r \end{bmatrix}, \quad \mathbf{k} = \begin{bmatrix} \mathbf{k}_m \\ \mathbf{k}_r \end{bmatrix}.$$

Using truncated observations, we can predict the value z_* as

$$\mu_{z_*|\mathbf{y}_m} = \mathbf{k}_m^T \mathbf{C}_m^{-1} \mathbf{k}_m, \tag{4.4}$$

with a prediction error variance given by

$$\sigma^2_{z_*|\mathbf{y}_m} = \sigma_f^2 (1 - \mathbf{k}_m^T \mathbf{C}_m^{-1} \mathbf{k}_m), \tag{4.5}$$

where $\mathbf{C}_m = \mathbf{K}_m + \sigma_w^2 \mathbf{I} \in \mathbb{R}^{m \times m}$.

The following result shows the gap between predicted values using truncated measurements and all measurements.

Theorem 4.1 *Consider a Gaussian process $z(\mathbf{x}) \sim \mathcal{GP}(0, \sigma_f^2 C(\mathbf{x}, \mathbf{x}'))$, we have*

$$\mu_{z_*|\mathbf{y}} - \mu_{z_*|\mathbf{y}_m} = (\mathbf{k}_r - \mathbf{K}_{mr}^T \mathbf{C}_m^{-1} \mathbf{k}_m)^T (\mathbf{C}_r - \mathbf{K}_{mr}^T \mathbf{C}_m^{-1} \mathbf{K}_{mr})^{-1} (\mathbf{y}_r - \mathbf{K}_{mr}^T \mathbf{C}_m^{-1} \mathbf{y}_m), \tag{4.6a}$$

and

$$\sigma^2_{z_*|y\mathbf{y}} - \sigma^2_{z_*|\mathbf{y}_m} = -\sigma_f^2 (\mathbf{k}_r - \mathbf{K}_{mr}^T \mathbf{C}_m^{-1} \mathbf{k}_m)^T (C_r - \mathbf{K}_{mr}^T \mathbf{C}_m^{-1} K_{mr})^{-1} (\mathbf{k}_r - \mathbf{K}_{mr}^T \mathbf{C}_m^{-1} \mathbf{k}_m) < 0. \tag{4.6b}$$

Proof We can rewrite (4.3a) as

$$\mu_{z_*|\mathbf{y}} = \begin{bmatrix} \mathbf{k}_m \\ \mathbf{k}_r \end{bmatrix}^T \begin{bmatrix} \mathbf{C}_m & \mathbf{K}_{mr} \\ \mathbf{K}_{mr}^T & \mathbf{C}_r \end{bmatrix}^{-1} \begin{bmatrix} \mathbf{y}_m \\ \mathbf{y}_r \end{bmatrix}, \tag{4.7a}$$

and (4.3b) as

$$\sigma^2_{z_*|\mathbf{y}} = \sigma_f^2 \left(1 - \begin{bmatrix} \mathbf{k}_m \\ \mathbf{k}_r \end{bmatrix}^T \begin{bmatrix} \mathbf{C}_m & \mathbf{K}_{mr} \\ \mathbf{K}_{mr}^T & \mathbf{C}_r \end{bmatrix}^{-1} \begin{bmatrix} \mathbf{k}_m \\ \mathbf{k}_r \end{bmatrix} \right). \tag{4.7b}$$

Using the identity based on matrix inversion lemma (see Appendix A.2), (4.7a) and (4.7b) become

$$\begin{aligned} \mu_{z_*|\mathbf{y}} &= \mathbf{k}_m^T \mathbf{C}_m^{-1} \mathbf{y}_m \\ &\quad + (\mathbf{k}_r - \mathbf{K}_{mr}^T \mathbf{C}_m^{-1} \mathbf{k}_m)^T (\mathbf{C}_r - \mathbf{K}_{mr}^T \mathbf{C}_m^{-1} \mathbf{K}_{mr})^{-1} (\mathbf{y}_r - \mathbf{K}_{mr}^T \mathbf{C}_m^{-1} \mathbf{y}_m), \end{aligned}$$

and

$$\begin{aligned} \sigma^2_{z_*|\mathbf{y}} &= \sigma_f^2 \left(1 - \mathbf{k}_m^T \mathbf{C}_m^{-1} \mathbf{k}_m\right) \\ &\quad - \sigma_f^2 (\mathbf{k}_r - \mathbf{K}_{mr}^T \mathbf{C}_m^{-1} \mathbf{k}_m)^T (\mathbf{C}_r - \mathbf{K}_{mr}^T \mathbf{C}_m^{-1} \mathbf{K}_{mr})^{-1} (\mathbf{k}_r - \mathbf{K}_{mr}^T \mathbf{C}_m^{-1} \mathbf{k}_m). \end{aligned}$$

Hence, by the use of (4.4) and (4.5), we obtain (4.6a) and 4.6b. □

Corollary 4.1 *The prediction error variance* $\sigma^2_{z_*|\mathbf{y}_m}$ *is a non-increasing function of* m.

Proof The proof is straightforward from Theorem 4.1 by letting $n = m + 1$. □

Considering an ideal case in which the measurements y_m are not correlated with the remaining measurements y_r, we have the following result.

Proposition 4.1 *Under the assumptions used in Theorem 4.1 and for given* $\mathbf{y}_r \sim \mathbb{N}(0, \mathbf{C}_r)$, *if* $\mathbf{K}_{mr} = 0$, *then* $\mu_{z_*|\mathbf{y}} - \mu_{z_*|\mathbf{y}_m} = \mathbf{k}_r^T \mathbf{C}_r^{-1} \mathbf{y}_r$ *and* $\sigma^2_{z_*|\mathbf{y}} - \sigma^2_{z_*|\mathbf{y}_m} = -\sigma_f^2 \mathbf{k}_r^T \mathbf{C}_r^{-1} \mathbf{k}_r$. *In addition, we also have*

$$\left|\mu_{z_*|\mathbf{y}} - \mu_{z_*|\mathbf{y}_m}\right| \leq \left\| \mathbf{k}_r^T \mathbf{C}_r^{-1} \right\| \sqrt{r}\, \bar{y}(p_1)$$

with a non-zero probability p_1. *For a desired* p_1, *we can find* $\bar{y}(p_1)$ *by solving*

$$p_1 = \prod_{1 \leq i \leq r} \left(1 - 2\Phi\left(-\frac{\bar{y}(p_1)}{\lambda_i^{1/2}} \right) \right), \tag{4.8}$$

where Φ is the cumulative normal distribution and $\{\lambda_i \mid i = 1, \ldots, r\}$ are the eigenvalues of $\mathbf{C}_r = \mathbf{U}\boldsymbol{\Lambda}\mathbf{U}^T$ with a unitary matrix $\mathbf{U}$, and $\boldsymbol{\Lambda} = \mathrm{diag}(\lambda_1, \ldots, \lambda_r)$.

Proof The first statement is straightforward from Theorem 4.1.

For the second statement, we can represent $\mathbf{y}_r$ as $\mathbf{y}_r = \mathbf{C}_r^{1/2}\mathbf{u} = \mathbf{U}\boldsymbol{\Lambda}^{1/2}\mathbf{u} = \mathbf{U}\tilde{\mathbf{y}}$, where $\mathbf{u}$ is a vector of independent standard normals and $\mathbf{C}_r = \mathbf{U}\boldsymbol{\Lambda}\mathbf{U}^T$ and $\mathbf{C}_r^{1/2} = \mathbf{U}\boldsymbol{\Lambda}^{1/2}$. By using the Cauchy-Schwarz inequality and norm inequalities, we have

$$
\begin{aligned}
\left|\mu_{z_*|\mathbf{y}} - \mu_{z_*|\mathbf{y}_m}\right| &= \left|\mathbf{k}_r^T\mathbf{C}_r^{-1}\mathbf{y}_r\right| = \left|\mathbf{k}_r^T\mathbf{C}_r^{-1}\mathbf{U}\tilde{\mathbf{y}}\right| \\
&\leq \left\|\mathbf{k}_r^T\mathbf{C}_r^{-1}\right\| \left\|\mathbf{U}\tilde{\mathbf{y}}\right\| = \left\|\mathbf{k}_r^T\mathbf{C}_r^{-1}\right\| \left\|\tilde{\mathbf{y}}\right\| \\
&\leq \left\|\mathbf{k}_r^T\mathbf{C}_r^{-1}\right\| \sqrt{r} \left\|\tilde{\mathbf{y}}\right\|_\infty \leq \left\|\mathbf{k}_r^T\mathbf{C}_r^{-1}\right\| \sqrt{r}\bar{y}.
\end{aligned}
$$

Recall that we have $\mathbf{u} \sim \mathbb{N}(0, \mathbf{I})$ and $\tilde{\mathbf{y}} \sim \mathbb{N}(0, \boldsymbol{\Lambda})$, where $\boldsymbol{\Lambda} = \mathrm{diag}(\lambda_1, \ldots, \lambda_r)$. Then we can compute the probability $p_1 = \Pr(\left\|\tilde{\mathbf{y}}\right\|_\infty \leq \bar{y})$ as follows.

$$
\begin{aligned}
p_1 &= \Pr\left(\max_{1\leq i\leq r} \left|\tilde{y}^{(i)}\right| \leq \bar{y}\right) = \Pr\left(\max_{1\leq i\leq r} \left|\lambda_i^{1/2} u_i\right| \leq \bar{y}\right) \\
&= \prod_{1\leq i\leq r} \Pr\left(\lambda_i^{1/2}|u_i| \leq \bar{y}\right) = \prod_{1\leq i\leq r} \Pr\left(|u_i| \leq \frac{\bar{y}}{\lambda_i^{1/2}}\right) \\
&= \prod_{1\leq i\leq r} \left(1 - 2\Phi\left(-\frac{\bar{y}}{\lambda_i^{1/2}}\right)\right),
\end{aligned}
$$

where $\Phi(\cdot)$ is the cumulative standard normal distribution. □

Hence, if the magnitude of $\mathbf{K}_{mr}$ is small, then the truncation error from using truncated measurements will be close to $\mathbf{k}_r^T\mathbf{C}_r^{-1}\mathbf{k}_r$. Furthermore, if we want to reduce this error, we want $\mathbf{k}_r$ to be small, i.e., when the covariance between z_* and the remaining measurements $\mathbf{y}_r$ is small. In summary, if the following two conditions are satisfied: (1) the correlation between measurements $\mathbf{y}_m$ and the remaining measurements $\mathbf{y}_r$ is small and (2) the correlation between z_* and the remaining measurements $\mathbf{y}_r$ is small, then the truncation error is small and $\mu_{z_*|\mathbf{y}_m}$ can be a good approximation to $\mu_{z_*|\mathbf{y}}$. This idea is formalized in a more general setting in the following theorem.

Theorem 4.2 *Consider a zero-mean Gaussian process $z(\mathbf{x}) \sim \mathbb{N}(0, \sigma_f^2 C(\mathbf{x}, \mathbf{x}'))$ with the correlation function*

$$
C(\mathbf{x}, \mathbf{x}') = \exp\left\{-\frac{\left\|\mathbf{x} - \mathbf{x}'\right\|^2}{2\sigma_\ell^2}\right\}, \tag{4.9}
$$

and assume that we have collected n observations, $y^{(1)}, \ldots, y^{(n)}$*. Suppose that* $\mathbf{K}_{mr}$ *is small enough such that* $\left\|\mathbf{K}_{mr}^T\mathbf{C}_m^{-1}\mathbf{k}_m\right\| \leq \|\mathbf{k}_r\|$*, and* $\left\|\mathbf{K}_{mr}^T\mathbf{C}_m^{-1}\mathbf{y}_m\right\| \leq \delta_2 \left\|\mathbf{y}_r\right\|$ *and for some* $\delta_2 > 0$*. Given* $0 < p_2 < 1$*, choose* $\bar{y}(p_2)$ *such that* $\max_{i=m+1}^n \left|y^{(i)}\right| < \bar{y}(p_2)$ *with probability* p_2 *and* $\epsilon > 0$ *such that* $\epsilon < 2\gamma r(1+\delta_2)\bar{y}(p_2)$ *where* γ *is the signal-to-noise ratio. For* $\mathbf{x}_*$*, if the last* $r = n - m$ *data points satisfy*

$$\left\|\mathbf{x}^{(i)} - \mathbf{x}_*\right\|^2 > 2\sigma_\ell^2 \log\left(2\gamma\frac{1}{\epsilon}r(1+\delta_2)\bar{y}(p_2)\right),$$

then, with probability p_2*, we have*

$$\left|\mu_{z_*|\mathbf{y}} - \mu_{z_*|\mathbf{y}_m}\right| < \epsilon.$$

Proof Let $\mathbf{A} = \mathbf{C}_m^{-1}\mathbf{K}_{mr}$ and $\mathbf{B} = \mathbf{K}_{mr}^T\mathbf{C}_m^{-1}\mathbf{K}_{mr}$ for notational convenience. Then

$$\begin{aligned}
\left|\mu_{z_*|\mathbf{y}} - \mu_{z_*|\mathbf{y}_m}\right| &= \left\|(\mathbf{k}_r^T - \mathbf{k}_m^T\mathbf{A})(\mathbf{C}_r - \mathbf{B})^{-1}(\mathbf{k}_r - \mathbf{A}^T\mathbf{y}_m)\right\| \\
&\leq \left\|\mathbf{k}_r^T - \mathbf{k}_m^T\mathbf{A}\right\| \left\|(\mathbf{C}_r - \mathbf{B})^{-1}(\mathbf{y}_r - \mathbf{A}^T\mathbf{y}_m)\right\| \\
&\leq \left\|\mathbf{k}_r^T - \mathbf{k}_m^T\mathbf{A}\right\| \times (\left\|(\mathbf{C}_r - \mathbf{B})^{-1}\mathbf{y}_r\right\| + \left\|(\mathbf{C}_r - \mathbf{B})^{-1}\mathbf{A}^T\mathbf{y}_m\right\|) \\
&\leq 2\left\|\mathbf{k}_r\right\| \left(\left\|(\mathbf{C}_r - \mathbf{B})^{-1}\mathbf{y}_r\right\| + \left\|(\mathbf{C}_r - \mathbf{B})^{-1}\mathbf{A}^T\mathbf{y}_m\right\|\right)
\end{aligned}$$

Since $\mathbf{K}_r$ is positive semi-definite, and $\mathbf{C}_m$ is positive definite, we have $\mathbf{K}_r - \mathbf{B}$ is positive semi-definite. Then we have

$$(\mathbf{C}_r - \mathbf{B})^{-1} = (\mathbf{K}_r + 1/\gamma\mathbf{I} - \mathbf{B})^{-1} \preceq \gamma\mathbf{I}.$$

Combining this result, we get

$$\begin{aligned}
\left|\mu_{z_*|\mathbf{y}} - \mu_{z_*|\mathbf{y}_m}\right| &\leq 2\gamma\left\|\mathbf{k}_r\right\|(\left\|\mathbf{y}_r\right\| + \left\|\mathbf{A}^T\mathbf{y}_m\right\|) \\
&\leq 2\gamma(1+\delta_2)\left\|\mathbf{k}_r\right\|\left\|\mathbf{y}_r\right\| \\
&\leq 2\gamma(1+\delta_2)\sqrt{r}C_{\max}\left\|\mathbf{y}_r\right\|,
\end{aligned}$$

where $C(\mathbf{x}^{(i)}, \mathbf{x}_*) \leq C_{\max}$ for $i \in \{m+1, \ldots, n\}$. Define $\bar{y}(p_2)$ such that $\max_{i=m+1}^n \left|y^{(i)}\right| \leq \bar{y}(p_2)$ with probability p_2. Then, with probability p_2, we have

$$\left|\mu_{z_*|\mathbf{y}} - \mu_{z_*|\mathbf{y}_m}\right| \leq 2\gamma r(1+\delta_2)C_{\max}\bar{y}(p_2).$$

Hence, for $\epsilon > 0$, if

$$C_{\max} < \frac{\epsilon}{2\gamma r(1+\delta_2)\bar{y}(p_2)} \tag{4.10}$$

with probability p_2, we have

$$\left|\mu_{z_*|\mathbf{y}} - \mu_{z_*|\mathbf{y}_m}\right| < \epsilon.$$

Let $l^2 = \min \left\|\mathbf{x}^{(i)} - \mathbf{x}_*\right\|^2$ for any $i \in \{m+1, \ldots, n\}$. Then (4.10) becomes, with probability p_2,

$$\exp\left(-\frac{l^2}{2\sigma_\ell^2}\right) \le C_{\max} < \frac{\epsilon}{2\gamma r(1+\delta_2)\bar{y}(p_2)}$$

$$l^2 > -2\sigma_\ell^2 \log\left(\frac{\epsilon}{2\gamma r(1+\delta_2)\bar{y}(p_2)}\right)$$

For $\epsilon < 2\gamma r(1+\delta_2)\bar{y}(p_2)$, we have

$$l^2 > 2\sigma_\ell^2 \log\left(2\gamma\frac{1}{\epsilon}r(1+\delta_2)\bar{y}(p_2)\right),$$

and this completes the proof. □

Remark 4.1 The last part of Proposition 4.1 and Theorem 4.2 seek a bound for the difference between predicted values using all and truncated observations with a given probability since the difference is a random variable.

Example 4.1 We provide an illustrative example to show how to use the result of Theorem 4.2 as follows. Consider a Gaussian process defined in (4.1) and (4.9) with $\sigma_f^2 = 1$, $\sigma_\ell = 0.2$, and $\gamma = 100$. If we have any randomly chosen 10 samples ($m = 10$) within $(0, 1)^2$ and we want to make prediction at $\mathbf{x}_* = (1, 1)^T$. We choose $\bar{y}(p_2) = 2\sigma_f = 2$ such that $\max_{i=m+1}^n \left|y^{(i)}\right| < \bar{y}(p_2)$ with probability $p_2 = 0.95$. According to Theorem 4.2, if we have an extra sample $\mathbf{x}^{(11)}$ ($r = 1$) at $(2.5, 2.5)^T$, which satisfies the condition $\left\|\mathbf{x}^{(11)} - \mathbf{x}_*\right\| > 0.92$, then the difference in prediction using with and without the extra sample is less than $\epsilon = 0.01$ with probability $p_2 = 0.95$.

Example 4.2 Motivated by the results presented, we take a closer look at the usefulness of using a subset of observations from a sensor network for a particular realization of the Gaussian process. We consider a particular realization shown in Fig. 4.2, where crosses represent the sampling points of a Gaussian process defined in (4.1) and (4.9) with $\sigma_f^2 = 1$, $\sigma_\ell = 0.2$, and $\gamma = 100$ over $(0, 1)^2$. We have selected y_m as the collection of observations (blue crosses) within the red circle of a radius $R = 2\sigma_\ell = 0.4$ centered at a point (a red star) located at $\mathbf{x}_* = (0.6, 0.4)^T$. If a measurement is taken outside the red circle, the correlation between this measurement and the value at $\mathbf{x}_*$ decreases to 0.135. The rest of observations (blue crosses outside

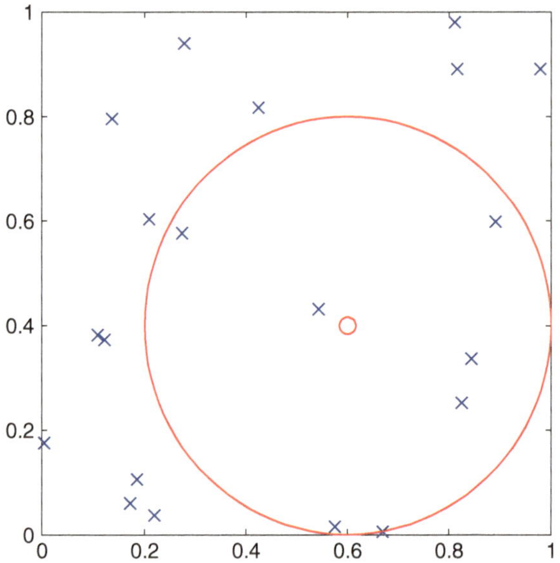

Fig. 4.2 Example of the selection of truncated observations. The parameters used in the example are: $\sigma_f^2 = 1$, $\sigma_\ell = 0.2$, $\sigma_w = 0.1$

of the red circle) are selected as y_r. The prediction results are shown in Table 4.1. In this particular realization, we have $z_* = 1.0298$. It can be seen that the prediction means and variances using only $\mathbf{y}_m$ are close to the one using all observations. We also compute the prediction at $\mathbf{x}_*$ with $\mathbf{y}_r$ which is far from the true value with a large variance.

The result of Theorem 4.2 and Examples 4.1 and 4.2 all suggest the usage of observations that are highly correlated with the point of interest.

4.1.2 Selecting Temporal Truncation Size

In previous subsection, we have obtained the error bounds for the prediction at a single point. In general, the observations made close to that point are more informative than the others.

Table 4.1 Prediction means and variances using $\mathbf{y}$, $\mathbf{y}_m$, and $\mathbf{y}_r$

	$n = 20$	$m = 12$	$r = 8$
$\mu_{z_*\|\mathbf{y}}$	1.0515	1.0633	0.3491
$\sigma^2_{z_*\|\mathbf{y}}$	0.0079	0.0080	0.9364

Consider a zero-mean spatio-temporal Gaussian process

$$z(\mathbf{s}, t) \sim \mathcal{GP}(0, \sigma_f^2 C(\mathbf{s}, t, \mathbf{s}', t')), \tag{4.11}$$

with covariance function

$$\begin{aligned} C(\mathbf{x}, \mathbf{x}') &= C_s(\mathbf{s}, \mathbf{s}')C_t(t, t') \\ &= \exp\left(-\sum_{\ell=1,2} \frac{(s_\ell - s'_\ell)^2}{2\sigma_\ell^2}\right) \exp\left(-\frac{(t-t')^2}{2\sigma_t^2}\right). \end{aligned} \tag{4.12}$$

We define η as the truncation size, and our objective is to use only the observations made during the last η time steps, i.e., from time $t-\eta+1$ to time t, to make prediction at time t. In general, a small η yields faster computation but lower accuracy and a large η yields slower computation but higher accuracy. Thus, the truncation size η should be selected according to a trade-off relationship between accuracy and efficiency.

Next, we show an approach to select the truncation size η in an averaged performance sense. Given the observations and associated sampling locations and times (denoted by $\mathcal{D}$ which depends on η), the generalization error $\epsilon_{\mathbf{x}_*,\mathcal{D}}$ at a point $\mathbf{x}_* = (\mathbf{s}_*^T, t_*)^T$ is defined as the prediction error variance $\sigma^2_{z_*|\mathcal{D}}$ [102, 103]. For a given t_* not knowing user specific $\mathbf{s}_*$ *a priori*, we seek to find η that guarantees a low prediction error variance uniformly over the entire space $\mathcal{Q}$, i.e., we want $\epsilon_{\mathcal{D}} = \mathbb{E}_{\mathbf{s}_*}[\sigma^2_{z_*|\mathcal{D}}]$ to be small [102, 103]. Here $\mathbb{E}_{\mathbf{s}_*}$ denotes the expectation with respect to the uniform distribution of $\mathbf{s}_*$.

According to Mercer's Theorem, we know that the kernel function C_s can be decomposed into

$$C_s(\mathbf{s}, \mathbf{s}') = \sum_{i=1}^{\infty} \lambda_i \phi_i(\mathbf{s}) \phi_i(\mathbf{s}'),$$

where $\{\lambda_i\}$ and $\{\phi_i(\cdot)\}$ are the eigenvalues and corresponding eigenfunctions, respectively [103]. In a similar way shown in [103], the input dependent generalization error $\epsilon_{\mathcal{D}}$ for our spatio-temporal Gaussian process can be obtained as

$$\begin{aligned} \epsilon_{\mathcal{D}} &= \mathbb{E}_{\mathbf{s}_*}\left[\sigma_f^2\left(1 - \operatorname{tr}\left(\mathbf{k}\mathbf{k}^T(\mathbf{K} + 1/\gamma\mathbf{I})^{-1}\right)\right)\right] \\ &= \sigma_f^2\left(1 - \operatorname{tr}\left(\mathbb{E}_{\mathbf{s}_*}[\mathbf{k}\mathbf{k}^T](\mathbf{K} + 1/\gamma\mathbf{I})^{-1}\right)\right). \end{aligned} \tag{4.13}$$

We have

$$\mathbb{E}_{\mathbf{s}_*}[\mathbf{k}\mathbf{k}^T] = \mathbf{\Psi}\mathbf{\Lambda}^2\mathbf{\Psi}^T \circ \mathbf{k}_t\mathbf{k}_t^T, \tag{4.14}$$

and

$$\mathbf{K} = \mathbf{\Psi}\mathbf{\Lambda}\mathbf{\Psi}^T \circ \mathbf{K}_t\mathbf{K}_t^T, \tag{4.15}$$

where $(\boldsymbol{\Psi})_{ij} = \phi_j(\mathbf{s}_i)$, $(\mathbf{k}_t)_j = C_t(t^{(j)}, t_*)$, $(\mathbf{K}_t)_{ij} = C_t(t^{(i)}, t^{(j)})$, and $(\boldsymbol{\Lambda})_{ij} = \lambda_i \delta_{ij}$. δ_{ij} denotes the Dirac delta function. $\circ$ denotes the Hadamard (element-wise) product [103]. Hence, the input-dependent generalization error $\epsilon_{\mathcal{D}}$ can be computed analytically by plugging (4.14) and (4.15) into (4.13). Notice that $\epsilon_{\mathcal{D}}$ is a function of inputs (i.e., the sampling locations and times). To obtain an averaged performance level without the knowledge of the algorithmic sampling strategy *a priori*, we use an appropriate sampling distribution which models the stochastic behavior of the sampling strategy. Thus, further averaging over the observation set $\mathcal{D}$ with the samp-ing distribution yields $\epsilon(\eta) = \mathbb{E}_{\mathcal{D}}[\epsilon_{\mathcal{D}}]$ which is a function of the truncation size η only. This averaging process can be done using Monte Carlo methods. Then η can be chosen based on the averaged performance measure $\epsilon(\eta)$ under the sampling distribution.

An alternative way, without using the eigenvalues and eigenfunctions, is to directly and numerically compute $\epsilon_{\mathcal{D}} = \mathbb{E}_{\mathbf{s}_*}[\sigma^2_{z_*|\mathcal{D}}]$ uniformly over the entire space $\mathcal{Q}$ with random sampling positions at each time step. An averaged generalization error with respect to the temporal truncation size can be plotted by using such Monte Carlo methods. Then the temporal truncation size η can be chosen such that a given level of the averaged generalization error is achieved.

Example 4.3 Consider a problem of selecting a temporal truncation size η for spatio- temporal Gaussian process regression using observations from 9 agents. The spatio-temporal Gaussian process is defined in (4.1) and (4.9) with $\sigma_f^2 = 1$, $\sigma_1 = \sigma_2 = 0.2$, $\sigma_t = 5$, and $\gamma = 100$ over $(0, 1)^2$. The Monte Carlo simulation result is shown in Fig. 4.3. The achieved generalization error $\epsilon_{\mathcal{D}}$ are plotted in blue circles with error-bars with respect to the temporal truncation size η. As can be seen, an averaged generalization error (in blue circles) under 0.1 can be achieved by using observations taken from last 10 time steps.

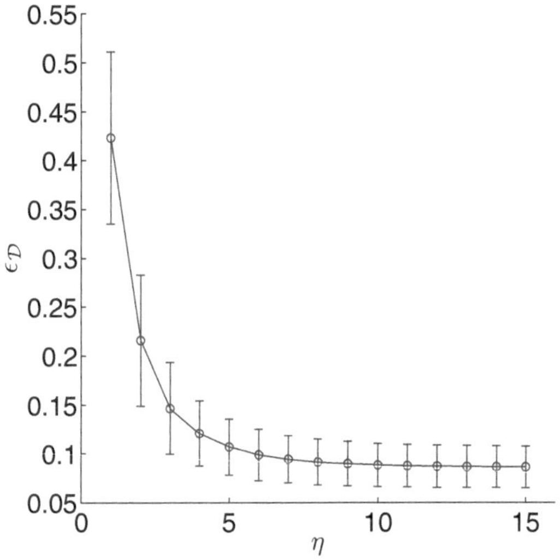

Fig. 4.3 Example of selecting a temporal truncation size η. The parameters used in the example are: $\sigma_f^2 = 1$, $\sigma_1 = \sigma_2 = 0.2$, $\sigma_t = 5$, $\gamma = 100$

Notice that the prediction error variances can be significantly minimized by optimally selecting the sampling positions. Hence, the selected η guarantees at least the averaged performance level of the sensor network when the optimal sampling strategy is used.

By using a fixed truncation size η, the computational complexity and memory space required for making prediction (i.e., evaluating (4.3a) and (4.3b)) do not increase as more measurements are collected. Our next objective is to improve the quality of the prediction by carefully selecting the future sampling positions for the mobile sensor network.

4.2 Optimal Sampling Strategies

At time t, the goal of the mobile sensor network is to make prediction at pre-specified points of interest $\{\mathbf{p}_j = (\mathbf{v}_j, \tau_j) \mid j \in \mathcal{J}\}$ indexed by $\mathcal{J} := \{1, \ldots, M\}$. From here on, points of interest will be referred to as *target points*. The introduction of target points is motivated by the fact that the potential environmental concerns should be frequently monitored. For instance, the target points can be assigned at the interface of a factory and a lake, sewage systems, or polluted beaches. Thus, the introduction of target points, which can be arbitrarily specified by a user, provides a flexible way to define a geometrical shape of a subregion of interest in a surveillance region. Notice that the target points can be changed by a user at any time. In particular, we allow that the number of target points M can be larger than that of agents N, which is often the case in practice. The prediction of $z_j := z(\mathbf{p}_j)$ of the Gaussian process at a target point $\mathbf{p}_j$ can be obtained as in (4.3a) and (4.3b).

4.2.1 Centralized Navigation Strategy

Consider the case in which a central station receives collective measurements from all N mobile sensors and performs the prediction. Let the central station discard the oldest set of measurements $\mathbf{y}_{t-\eta+1}$ after making the prediction at time t. At the next time index $t+1$, using the remained observations $\mathbf{y}_{t-\eta+2:t}$ in the memory along with new measurements $\mathbf{y}_{t+1}$ from all N agents at time $t+1$, the central station will predict $z(\mathbf{s}_*, t_*)$ evaluated at target points $\{\mathbf{p}_j \mid j \in \mathcal{J}\}$. Hence, agents should move to the most informative locations for taking measurements at time $t+1$ [44].

For notational simplicity, let $\bar{\mathbf{y}} \in \mathbb{R}^{N(\eta-1)}$ be the remaining observations, i.e., $\bar{\mathbf{y}} := \mathbf{y}_{t-\eta+2:t}$, and $\tilde{\mathbf{y}} \in \mathbb{R}^N$ be the measurements that will be taken at positions $\tilde{\mathbf{q}} = (\tilde{\mathbf{q}}_1^T, \ldots, \tilde{\mathbf{q}}_N^T)^T \in \mathcal{Q}^N$ and time $t+1$. In contrast to the information-theoretic control strategies using the conditional entropy or the mutual information criterion [44, 72], in this chapter, the mobility of the robotic sensors will be designed such that they directly minimize the average of the prediction error variances over target points, i.e.,

$$J_c(\tilde{\mathbf{q}}) = \frac{1}{|\mathcal{J}|} \sum_{j \in \mathcal{J}} \sigma^2_{z_j|\bar{\mathbf{y}},\tilde{\mathbf{y}}}(\tilde{\mathbf{q}}), \tag{4.16}$$

where $|\mathcal{J}| = M$ is the cardinality of $\mathcal{J}$. The prediction error variance at each of M target points is given by

$$\sigma^2_{z_j|\bar{\mathbf{y}},\tilde{\mathbf{y}}}(\tilde{\mathbf{q}}) = \sigma_f^2 \left(1 - \mathbf{k}_j(\tilde{\mathbf{q}})^T \mathbf{C}(\tilde{\mathbf{q}})^{-1} \mathbf{k}_j(\tilde{\mathbf{q}})\right), \quad \forall j \in \mathcal{J},$$

where $\mathbf{k}_j(\tilde{\mathbf{q}})$ and $\mathbf{C}(\tilde{\mathbf{q}})$ are defined as

$$\mathbf{k}_j(\tilde{\mathbf{q}}) = \begin{bmatrix} \mathrm{Corr}(\bar{\mathbf{y}}, z_j) \\ \mathrm{Corr}(\tilde{\mathbf{y}}, z_j) \end{bmatrix}, \quad \mathbf{C}(\tilde{\mathbf{q}}) = \begin{bmatrix} \mathrm{Corr}(\bar{\mathbf{y}}, \bar{\mathbf{y}}) & \mathrm{Corr}(\bar{\mathbf{y}}, \tilde{\mathbf{y}}) \\ \mathrm{Corr}(\tilde{\mathbf{y}}, \bar{\mathbf{y}}) & \mathrm{Corr}(\tilde{\mathbf{y}}, \tilde{\mathbf{y}}) \end{bmatrix}.$$

In order to reduce the average of prediction error variances over target points $\{\mathbf{p}_j \mid j \in \mathcal{J}\}$, the central station solves the following optimization problem

$$\mathbf{q}(t+1) = \arg \min_{\tilde{\mathbf{q}} \in \mathcal{Q}^N} J_c(\tilde{\mathbf{q}}). \tag{4.17}$$

Notice that in this problem set-up, we only consider the constraint that robots should move within the region $\mathcal{Q}$. However, the mobility constraints such as the maximum distance a robot can move between two time indices or the maximum speed a robot can travel, can be incorporated as additional constraints in the optimization problem [45].

The sensor network configuration $\mathbf{q}(t)$ can be controlled by a gradient descent algorithm such that $\mathbf{q}(t)$ can move to a local minimum of J_c for the prediction at time $t+1$. The gradient descent control algorithm is given by

$$\frac{d\mathbf{q}(\tau)}{d\tau} = -\nabla_{\mathbf{q}} J_c(\mathbf{q}(\tau)), \tag{4.18}$$

where $\nabla_{\mathbf{x}} J_c(\mathbf{x})$ denotes the gradient of $J_c(\mathbf{x})$ at $\mathbf{x}$. A critical point of $J_c(\mathbf{q})$ obtained in (4.18) will be $\mathbf{q}(t+1)$. The analytical form of $\partial\sigma^2_{z_j|\bar{\mathbf{y}},\tilde{\mathbf{y}}}(\tilde{\mathbf{q}})/\partial\tilde{q}_{i,\ell}$, where $\tilde{q}_{i,\ell}$ is the ℓth element in $\tilde{\mathbf{q}}_i \in \mathcal{Q}$, can be obtained by

$$\frac{\partial\sigma^2_{z_j|\bar{\mathbf{y}},\tilde{\mathbf{y}}}(\tilde{\mathbf{q}})}{\partial\tilde{q}_{i,\ell}} = \mathbf{k}_j^T \mathbf{C}^{-1} \left(\frac{\partial\mathbf{C}}{\partial\tilde{q}_{i,\ell}} \mathbf{C}^{-1} \mathbf{k}_j - 2\frac{\partial\mathbf{k}_j}{\partial\tilde{q}_{i,\ell}}\right), \quad \forall i \in \mathcal{I}, \ell \in \{1, 2\}.$$

Other more advanced non-linear optimization techniques may be applied to solve the optimization problem in (4.17) [104].

The centralized sampling strategy for the mobile sensor network with the cost function J_c in (4.16) is summarized in Table 4.2. Notice that the prediction in the centralized sampling strategy uses temporally truncated observations. A decentralized version of the centralized sampling strategy in Table 4.2 may be developed using the approach proposed in [105] in which each robot incrementally refines its decision while intermittently communicating with the rest of the robots.

4.2.2 Distributed Navigation Strategy

Now, we consider a case in which each agent in the sensor network can only communicate with other agents within a limited communication range R. In addition, no central station exists. In this section, we present a distributed navigation strategy for mobile agents that uses only local information in order to minimize a collective network performance cost function.

The communication network of mobile agents can be represented by an undirected graph. Let $\mathcal{G}(t) := (\mathcal{I}, \mathcal{E}(t))$ be an undirected communication graph such that an edge $(i, j) \in \mathcal{E}(t)$ if and only if agent i can communicate with agent j at time t. We define the neighborhood of agent i at time t by $\mathcal{N}_i(t) := \{j \in \mathcal{I} \mid (i, j) \in \mathcal{E}(t)\}$. In particular, we have

$$\mathcal{N}_i(t) = \left\{j \in \mathcal{I} \mid \left\|\mathbf{q}_i(t) - \mathbf{q}_j(t)\right\| < R, j \neq i\right\}.$$

Note that in our definition above, "$<$" is used instead of "$\leq$" in deciding the communication range.

At time $t \in \mathbb{Z}_{>0}$, agent i collects measurements $\left\{y_j(t) \mid j \in \{i\} \cup \mathcal{N}_i(t)\right\}$ sampled at $\left\{\mathbf{q}_j(t) \mid j \in \{i\} \cup \mathcal{N}_i(t)\right\}$ from its neighbors and itself. The collection of these observations and the associated sampling positions in vector forms are denoted by $\mathbf{y}_t^{[i]}$ and $\mathbf{q}_t^{[i]}$, respectively. Similarly, for notational simplicity, we also define the cumulative measurements that have been collected by agent i from time $t - \eta + 1$ to t as

$$\mathbf{y}_{t-\eta+1:t}^{[i]} = \left((\mathbf{y}_{t-\eta+1}^{[i]})^T, \ldots, (\mathbf{y}_t^{[i]})^T\right)^T.$$

In contrast to the centralized scheme, in the distributed scheme, each agent determines the sampling points based on the local information from neighbors. After making the prediction at time t, agent i discards the oldest set of measurements $\mathbf{y}_{t-\eta+1}^{[i]}$. At time $t + 1$, using the remained observations $\mathbf{y}_{t-\eta+2:t}^{[i]}$ in the memory along with new measurements $\mathbf{y}_{t+1}^{[i]}$ from its neighbors in $\mathcal{N}_i(t + 1)$, agent i will predict $z(\mathbf{s}_*, t_*)$ evaluated at target points $\left\{\mathbf{p}_j \mid j \in \mathcal{J}\right\}$.

Table 4.2 Centralized sampling strategy at time t

Input:
(1) Number of agents N
(2) Positions of agents $\{\mathbf{q}_i(t) \mid i \in \mathcal{I}\}$
(3) Hyperparameters of the Gaussian process $\boldsymbol{\theta} = (\sigma_f^2, \sigma_1, \sigma_2, \sigma_t)^T$
(4) Target points $\{\mathbf{p}_j \mid j \in \mathcal{J}\}$
(5) Truncation size η
Output:
(1) Prediction at target points $\left\{\mu_{z_j \mid \mathbf{y}_{t-\eta+1:t}} \mid j \in \mathcal{J}\right\}$
(2) Prediction error variance at target points $\left\{\sigma^2_{z_j \mid \mathbf{y}_{t-\eta+1:t}} \mid j \in \mathcal{J}\right\}$

For $i \in \mathcal{I}$, agent i performs:
make an observation at current position $\mathbf{q}_i(t)$, *i.e.*, $y_i(t)$
transmit the observation $y_i(t)$ to the central station

The central station performs:
collect the observations from all N agents, *i.e.*, $\mathbf{y}_t = (y_1(t), \cdots, y_N(t))^T$
obtain the cumulative measurements, *i.e.*, $\mathbf{y}_{t-\eta+1:t} = (\mathbf{y}_{t-\eta+1}^T, \cdots, \mathbf{y}_t^T)^T$
for $j \in \mathcal{J}$ **do**
  make prediction at a target point $\mathbf{p}_j$
$$\mu_{z_j \mid \mathbf{y}_{t-\eta+1:t}} = \mathbf{k}^T \mathbf{C}^{-1} \mathbf{y},$$
  with a prediction error variance given by
$$\sigma^2_{z_j \mid \mathbf{y}_{t-\eta+1:t}} = \sigma_f^2 (1 - \mathbf{k}^T \mathbf{C}^{-1} \mathbf{k}),$$
  where $\mathbf{y} = \mathbf{y}_{t-\eta+1:t}$, $\mathbf{k} = \mathrm{Corr}(\mathbf{y}, z_j)$, and $\mathbf{C} = \mathrm{Corr}(\mathbf{y}, \mathbf{y})$
end for
if $t \geq \eta$ **then**
  discard the oldest set of measurements taken at time $t - \eta + 1$, *i.e.*, $\mathbf{y}_{t-\eta+1}$
end if
compute the control with the remained data $\mathbf{y}_{t-\eta+2:t}$
$$\mathbf{q}(t+1) = \arg\min_{\tilde{\mathbf{q}} \in \mathcal{Q}^N} J_c(\tilde{\mathbf{q}}),$$
via
$$\frac{d\mathbf{q}(\tau)}{d\tau} = -\nabla_{\mathbf{q}} J_c(\mathbf{q}(\tau))$$
send the next sampling positions $\{\mathbf{q}_i(t+1)\}_{i=1}^N$ (a critical point of $J_c(\tilde{\mathbf{q}})$) to all N agents

For $i \in \mathcal{I}$, agent i performs:
receive the next sampling position $\mathbf{q}_i(t+1)$ from the central station
move to $\mathbf{q}_i(t+1)$ before time $t+1$

For notational simplicity, let $\bar{\mathbf{y}}^{[i]}$ be the remaining observations of agent i, i.e., $\bar{\mathbf{y}}^{[i]} := \mathbf{y}^{[i]}_{t-\eta+2:t}$. Let $\tilde{\mathbf{y}}^{[i]}$ be the new measurements that will be taken at positions of agent i and its neighbors $\tilde{\mathbf{q}}^{[i]} \in \mathcal{Q}^{|\mathcal{N}_i(t+1)|+1}$, and at time $t+1$, where $|\mathcal{N}_i(t+1)|$ is the number of neighbors of agent i at time $t+1$. The prediction error variance obtained by agent i at each of M target points (indexed by $\mathcal{J}$) is given by

$$\sigma^2_{z_j|\bar{\mathbf{y}}^{[i]},\tilde{\mathbf{y}}^{[i]}}(\tilde{\mathbf{q}}^{[i]}) = \sigma_f^2\left(1 - \mathbf{k}_j^{[i]}(\tilde{\mathbf{q}}^{[i]})^T \mathbf{C}^{[i]}(\tilde{\mathbf{q}}^{[i]})^{-1}\mathbf{k}_j^{[i]}(\tilde{\mathbf{q}}^{[i]})\right), \quad \forall j \in \mathcal{J},$$

where $\mathbf{k}_j^{[i]}(\tilde{\mathbf{q}}^{[i]})$ and $\mathbf{C}^{[i]}(\tilde{\mathbf{q}}^{[i]})$ are defined as

$$\mathbf{k}_j^{[i]}(\tilde{\mathbf{q}}^{[i]}) = \begin{bmatrix} \mathrm{Corr}(\bar{\mathbf{y}}^{[i]}, z_j) \\ \mathrm{Corr}(\tilde{\mathbf{y}}^{[i]}, z_j) \end{bmatrix}, \quad \mathbf{C}^{[i]}(\tilde{\mathbf{q}}^{[i]}) = \begin{bmatrix} \mathrm{Corr}(\bar{\mathbf{y}}^{[i]}, \bar{\mathbf{y}}^{[i]}) & \mathrm{Corr}(\bar{\mathbf{y}}^{[i]}, \tilde{\mathbf{y}}^{[i]}) \\ \mathrm{Corr}(\tilde{\mathbf{y}}^{[i]}, \bar{\mathbf{y}}^{[i]}) & \mathrm{Corr}(\tilde{\mathbf{y}}^{[i]}, \tilde{\mathbf{y}}^{[i]}) \end{bmatrix}. \tag{4.19}$$

The performance of agent i can be evaluated by the average of the prediction error variances over target points, i.e.,

$$J^{[i]}(\tilde{\mathbf{q}}^{[i]}) = \frac{1}{|\mathcal{J}|}\sum_{j\in\mathcal{J}} \sigma^2_{z_j|\bar{\mathbf{y}}^{[i]},\tilde{\mathbf{y}}^{[i]}}(\tilde{\mathbf{q}}^{[i]}), \quad \forall i \in \mathcal{I}.$$

One criterion to evaluate the network performance is the average of individual performance, i.e.,

$$J(\tilde{\mathbf{q}}) = \frac{1}{|\mathcal{I}|}\sum_{i\in\mathcal{I}} J^{[i]}(\tilde{\mathbf{q}}^{[i]}). \tag{4.20}$$

However, the discontinuity of the function J occurs at the moment of gaining or losing neighbors, e.g., at the set

$$\left\{\tilde{\mathbf{q}} \mid \|\tilde{\mathbf{q}}_i - \tilde{\mathbf{q}}_j\| = R\right\}.$$

A gradient decent algorithm for mobile robots that minimizes such J may produce hybrid system dynamics and/or chattering behaviors when robots lose or gain neighbors.

Therefore, we seek to minimize an upper-bound of J that is continuously differentiable. Consider the following function

$$\bar{\sigma}^2_{z_j|\bar{\mathbf{y}}^{[i]},\tilde{\mathbf{y}}^{[i]}}(\tilde{\mathbf{q}}^{[i]}) = \sigma_f^2\left(1 - \mathbf{k}_j^{[i]}(\tilde{\mathbf{q}}^{[i]})^T \bar{\mathbf{C}}^{[i]}(\tilde{\mathbf{q}}^{[i]})^{-1}\mathbf{k}_j^{[i]}(\tilde{\mathbf{q}}^{[i]})\right), \quad \forall j \in \mathcal{J}, \tag{4.21}$$

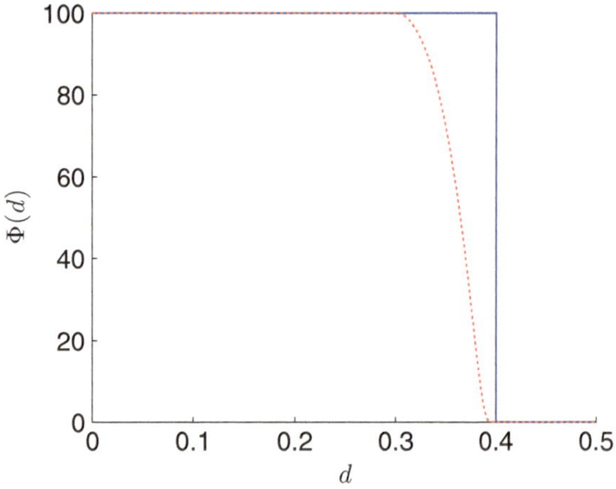

Fig. 4.4 Function $\Phi(d)$ in (4.22) with $\gamma = 100$, $R = 0.4$, and $d_0 = 0.1$ is shown in a *red dotted line*. The function $\Phi(d) = \gamma$ is shown in a *blue solid line*

where $\bar{\mathbf{C}}^{[i]}(\tilde{\mathbf{q}}^{[i]})$ is defined as

$$\bar{\mathbf{C}}^{[i]}(\tilde{\mathbf{q}}^{[i]}) = \begin{bmatrix} \mathrm{Corr}(\bar{\mathbf{y}}^{[i]}, \bar{\mathbf{y}}^{[i]}) & \mathrm{Corr}(\bar{\mathbf{y}}^{[i]}, \tilde{\mathbf{y}}^{[i]}) \\ \mathrm{Corr}(\tilde{\mathbf{y}}^{[i]}, \bar{\mathbf{y}}^{[i]}) & \mathrm{Corr}(\tilde{\mathbf{y}}^{[i]}, \tilde{\mathbf{y}}^{[i]}) + \tilde{\mathbf{C}}^{[i]}(\tilde{\mathbf{q}}^{[i]}) \end{bmatrix}.$$

Notice that $\bar{\mathbf{C}}^{[i]}(\tilde{\mathbf{q}}^{[i]})$ is obtained by adding a positive semi-definite matrix $\tilde{\mathbf{C}}^{[i]}(\tilde{\mathbf{q}}^{[i]})$ to the lower right block of $\mathbf{C}^{[i]}(\tilde{\mathbf{q}}^{[i]})$ in (4.19), where

$$\tilde{\mathbf{C}}^{[i]}(\tilde{\mathbf{q}}^{[i]}) = \mathrm{diag}\left(\Phi(d_{i1})^{-1}, \ldots, \Phi(d_{i(|\mathcal{N}_i(t+1)|+1)})^{-1}\right) - \frac{1}{\gamma}\mathbf{I},$$

where $d_{ij} := \left\|\tilde{\mathbf{q}}_i - \tilde{\mathbf{q}}_j\right\|$ is the distance between agent i and agent $j, \forall j \in \{i\} \cup \mathcal{N}_i(t+1)$. $\Phi : [0, R) \mapsto (0, \gamma]$ is a continuously differentiable function defined as

$$\Phi(d) = \gamma\phi\left(\frac{d + d_0 - R}{d_0}\right), \tag{4.22}$$

where

$$\phi(h) = \begin{cases} 1, & h \le 0, \\ \exp\left(\frac{-h^2}{1-h^2}\right), & 0 < h < 1. \end{cases}$$

An example of $\Phi(d)$ where $\gamma = 100$, $R = 0.4$, and $d_0 = 0.1$ is shown in the red dotted line in Fig. 4.4. Notice that if $\Phi(d) = \gamma$ is used (the blue solid line in Fig. 4.4), we have $\bar{\mathbf{C}}^{[i]}(\tilde{\mathbf{q}}^{[i]}) = \mathbf{C}^{[i]}(\tilde{\mathbf{q}}^{[i]})$. We then have the following result.

Proposition 4.2 *$\bar{\sigma}^2_{z_j|\bar{\mathbf{y}}^{[i]},\tilde{\mathbf{y}}^{[i]}}(\tilde{\mathbf{q}}^{[i]})$ is an upper-bound of $\sigma^2_{z_j|\bar{\mathbf{y}}^{[i]},\tilde{\mathbf{y}}^{[i]}}(\tilde{\mathbf{q}}^{[i]})$, $\forall i \in \mathcal{I}$.*

Proof Let $\mathbf{A} := \mathbf{C}^{[i]}(\tilde{\mathbf{q}}^{[i]})$ and $\mathbf{B} := \text{diag}(\mathbf{0}, \tilde{\mathbf{C}}^{[i]}(\tilde{\mathbf{q}}^{[i]}))$. The result follows immediately from the fact that $(\mathbf{A}+\mathbf{B})^{-1} \preceq \mathbf{A}^{-1}$ for any $\mathbf{A} \succ 0$ and $\mathbf{B} \succeq 0$. □

Hence, we construct a new cost function as

$$J_d(\tilde{\mathbf{q}}) = \frac{1}{|\mathcal{I}|}\sum_{i\in\mathcal{I}} \frac{1}{|\mathcal{J}|}\sum_{j\in\mathcal{J}} \bar{\sigma}^2_{z_j|\bar{\mathbf{y}}^{[i]},\tilde{\mathbf{y}}^{[i]}}(\tilde{\mathbf{q}}^{[i]}). \tag{4.23}$$

By Proposition 4.2, J_d in (4.23) is an upper-bound of J in (4.20).

Next, we show that J_d is continuously differentiable when agents gain or lose neighbors. In doing so, we compute the partial derivative of J_d with respect to $\tilde{\mathbf{q}}_{i,\ell}$, where $\tilde{\mathbf{q}}_{i,\ell}$ is the ℓth element in $\tilde{\mathbf{q}}_i \in \mathcal{Q}$, as follows.

$$\begin{aligned}
\frac{\partial J_d(\tilde{\mathbf{q}})}{\partial \tilde{q}_{i,\ell}} &= \frac{1}{|\mathcal{I}|}\sum_{k\in\mathcal{I}} \frac{1}{|\mathcal{J}|}\sum_{j\in\mathcal{J}} \frac{\partial \bar{\sigma}^2_{z_j|\bar{\mathbf{y}}^{[k]},\tilde{\mathbf{y}}^{[k]}}(\tilde{\mathbf{q}}^{[k]})}{\partial \tilde{q}_{i,\ell}} \\
&= \frac{1}{|\mathcal{I}|}\sum_{k\in\{i\}\cup\mathcal{N}_i} \frac{1}{|\mathcal{J}|}\sum_{j\in\mathcal{J}} \frac{\partial \bar{\sigma}^2_{z_j|\bar{\mathbf{y}}^{[k]},\tilde{\mathbf{y}}^{[k]}}(\tilde{\mathbf{q}}^{[k]})}{\partial \tilde{q}_{i,\ell}}, \quad \forall i \in \mathcal{I}, \ell \in \{1,2\}.
\end{aligned} \tag{4.24}$$

We then have the following.

Proposition 4.3 *The cost function J_d in (4.23) is of class $\mathcal{C}^1$, i.e., it is continuously differentiable.*

Proof We need to show that the partial derivatives of J_d with respect to $\tilde{q}_{i,\ell}$, $\forall i \in \mathcal{I}, \ell \in \{1,2\}$ exist and are continuous. Without loss of generality, we show that $\partial J_d/\partial \tilde{q}_{i,\ell}$, $\forall \ell \in \{1,2\}$ is continuous at any point $\tilde{\mathbf{q}}^*$ in the following boundary set defined by

$$\mathcal{S}_{ik} := \left\{\tilde{\mathbf{q}} \mid d_{ik} = \left\|\tilde{\mathbf{q}}_i - \tilde{\mathbf{q}}_k\right\| = R\right\}.$$

First, we consider a case in which $\tilde{\mathbf{q}} \notin \mathcal{S}_{ik}$ and $d_{ik} < R$, i.e., $k \in \mathcal{N}_i$ and $i \in \mathcal{N}_k$. By the construction of $\bar{\sigma}^2_{z_j|\bar{\mathbf{y}}^{[i]},\tilde{\mathbf{y}}^{[i]}}$ in (4.21) using (4.22), when we take the limit of the partial derivative, as d_{ik} approaches R from below (as $\tilde{\mathbf{q}}$ approaches $\tilde{\mathbf{q}}^*$), we have that

$$\begin{aligned}
\lim_{d_{ik}\to R_-} \frac{\partial \bar{\sigma}^2_{z_j|\bar{\mathbf{y}}^{[i]},\tilde{\mathbf{y}}^{[i]}}(\tilde{\mathbf{q}}^{[i]})}{\partial \tilde{q}_{i,\ell}} &= \frac{\partial \bar{\sigma}^2_{z_j|\bar{\mathbf{y}}^{[i]},\tilde{\mathbf{y}}^{[i]}}(\tilde{\mathbf{q}}^{[i]}\backslash\tilde{\mathbf{q}}_k)}{\partial \tilde{q}_{i,\ell}}, \\
\lim_{d_{ik}\to R_-} \frac{\partial \bar{\sigma}^2_{z_j|\bar{\mathbf{y}}^{[k]},\tilde{\mathbf{y}}^{[k]}}(\tilde{\mathbf{q}}^{[k]})}{\partial \tilde{q}_{i,\ell}} &= \frac{\partial \bar{\sigma}^2_{z_j|\bar{\mathbf{y}}^{[k]},\tilde{\mathbf{y}}^{[k]}}(\tilde{\mathbf{q}}^{[k]}\backslash\tilde{\mathbf{q}}_i)}{\partial \tilde{q}_{i,\ell}} = 0,
\end{aligned}$$

Table 4.3 Distributed sampling strategy at time t

Input:

(1) Number of agents N

(2) Positions of agents $\{\mathbf{q}_i(t) \mid i \in \mathcal{I}\}$

(3) Hyperparameters of the Gaussian process $\boldsymbol{\theta} = (\sigma_f^2, \sigma_1, \sigma_2, \sigma_t)^T$

(4) Target points $\{\mathbf{p}_j \mid j \in \mathcal{J}\}$

(5) Truncation size η

Output:

(1) Prediction at target points $\left\{\mu_{z_j \mid \mathbf{y}^{[i]}_{t-\eta+1:t}} \mid i \in \mathcal{I}, j \in \mathcal{J}\right\}$

(2) Prediction error variances at target points $\left\{\sigma^2_{z_j \mid \mathbf{y}^{[i]}_{t-\eta+1:t}} \mid i \in \mathcal{I}, j \in \mathcal{J}\right\}$

For $i \in \mathcal{I}$, agent i performs:

make an observation at $\mathbf{q}_i(t)$, *i.e.*, $y_i(t)$

transmit the observation to the neighbors in $\mathcal{N}_i(t)$

collect the observations from neighbors in $\mathcal{N}_i(t)$, *i.e.*, $\mathbf{y}^{[i]}(t)$

obtain the cumulative measurements, *i.e.*, $\mathbf{y}^{[i]}_{t-\eta+1:t} = \left((\mathbf{y}^{[i]}_{t-\eta+1})^T, \cdots, (\mathbf{y}^{[i]}_t)^T\right)^T$

for $j \in \mathcal{J}$ **do**

  make prediction at a target point $\mathbf{p}_j$

$$\mu_{z_j \mid \mathbf{y}^{[i]}_{t-\eta+1:t}} = \mathbf{k}^T \mathbf{C}^{-1} \mathbf{y},$$

  with a prediction error variance given by

$$\sigma^2_{z_j \mid \mathbf{y}^{[i]}_{t-\eta+1:t}} = \sigma_f^2 (1 - \mathbf{k}^T \mathbf{C}^{-1} \mathbf{k}),$$

  where $\mathbf{y} = \mathbf{y}^{[i]}_{t-\eta+1:t}$, $\mathbf{k} = \mathrm{Corr}(\mathbf{y}, z_j)$, and $\mathbf{C} = \mathrm{Corr}(\mathbf{y}, \mathbf{y})$

end for

if $t \geq \eta$ **then**

  discard the oldest set of measurements taken at time $t - \eta + 1$, *i.e.*, $\mathbf{y}^{[i]}_{t-\eta+1}$

end if

while $t \leq \tau \leq t + 1$ **do**

  compute $\nabla_{\mathbf{q}_\ell} J^{[i]}$ with the remained data $y^{[i]}_{t-\eta+2}$

  send $\nabla_{\mathbf{q}_\ell} J^{[i]}$ to agent ℓ in $\mathcal{N}_i(\tau)$

  receive $\nabla_{\mathbf{q}_i} J^{[\ell]}$ from all neighbors in $\mathcal{N}_i(\tau)$

  compute the gradient $\nabla_{\mathbf{q}_i} J_d = \sum_{\ell \in \mathcal{N}_i(\tau)} \nabla_{\mathbf{q}_i} J^{[\ell]} / |\mathcal{I}|$

  update position according to $\mathbf{q}_i(\tau + \delta t) = \mathbf{q}_i(\tau) - \alpha \nabla_{\mathbf{q}_i} J_d$ for a small step size α

end while

where $\tilde{\mathbf{q}}^{[a]} \backslash \tilde{\mathbf{q}}_b$ denotes the collection of locations of agent a and its neighbors excluding $\tilde{\mathbf{q}}_b$. Hence we have

$$\lim_{d_{ik} \to R_-} \frac{\partial J_d(\tilde{\mathbf{q}})}{\partial \tilde{q}_{i,\ell}} = \frac{\partial J_d(\tilde{\mathbf{q}}^*)}{\partial \tilde{q}_{i,\ell}}. \tag{4.25}$$

Consider the other case in which $\tilde{\mathbf{q}} \notin \mathcal{S}_{ik}$ and $d_{ik} > R$, i.e., $k \notin \mathcal{N}_i$ and $i \notin \mathcal{N}_k$. When d_{ik} approaches R from above (as $\tilde{\mathbf{q}}$ approaches $\tilde{\mathbf{q}}^*$), we have

$$\lim_{d_{ik} \to R_+} \frac{\partial \bar{\sigma}^2_{z_j | \bar{\mathbf{y}}^{[i]}, \tilde{\mathbf{y}}^{[i]}}(\tilde{\mathbf{q}}^{[i]})}{\partial \tilde{q}_{i,\ell}} = \frac{\partial \bar{\sigma}^2_{z_j | \bar{\mathbf{y}}^{[i]}, \tilde{\mathbf{y}}^{[i]}}(\tilde{\mathbf{q}}^{[i]})}{\partial \tilde{q}_{i,\ell}},$$

and hence

$$\lim_{d_{ik} \to R_+} \frac{\partial J_d(\tilde{\mathbf{q}})}{\partial \tilde{q}_{i,\ell}} = \frac{\partial J_d(\tilde{\mathbf{q}}^*)}{\partial \tilde{q}_{i,\ell}}. \tag{4.26}$$

Therefore, from (4.25) and (4.26), we have

$$\lim_{d_{ik} \to R_-} \frac{\partial J_d(\tilde{\mathbf{q}})}{\partial \tilde{q}_{i,\ell}} = \lim_{d_{ik} \to R_+} \frac{\partial J_d(\tilde{\mathbf{q}})}{\partial \tilde{q}_{i,\ell}} = \frac{\partial J_d(\tilde{\mathbf{q}}^*)}{\partial \tilde{q}_{i,\ell}}.$$

This completes the proof due to Theorem 4.6 in [106]. □

By using J_d in (4.23), a gradient descent algorithm can be used to minimize the network performance cost function J_d in (4.23) for the prediction at $t+1$.

$$\frac{d\mathbf{q}(\tau)}{d\tau} = -\nabla_{\mathbf{q}} J_d(\mathbf{q}(\tau)). \tag{4.27}$$

Note that the partial derivative in (4.24), which builds the gradient flow in (4.27), is a function of positions in $\cup_{j \in \mathcal{N}_i(t)} \mathcal{N}_j(t)$ only. This makes the algorithm distributed. A distributed sampling strategy for agent i with the network cost function J_d in (4.23) is summarized in Table 4.3. In this way, each agent with the distributed sampling strategy uses spatially and temporally truncated observations.

4.3 Simulation

In this section, we apply our approach to a spatio-temporal Gaussian process with a covariance function in (4.12). The Gaussian process was numerically generated through circulant embedding of the covariance matrix for the simulation [107]. The hyperparameters used in the simulation were chosen to be $\boldsymbol{\theta} = (\sigma_f^2, \sigma_1, \sigma_2, \sigma_t)^T = (1, 0.2, 0.2, 5)^T$. The surveillance region $\mathcal{Q}$ is given by $\mathcal{Q} = (0, 1)^2$. The signal-to-noise ratio $\gamma = 100$ is used throughout the simulation which is equivalent to a noise level of $\sigma_w = 0.1$. In our simulation, $N = 9$ agents sample at time $t \in \mathbb{Z}_{>0}$. The initial positions of the agents are randomly selected. The truncation size $\eta = 10$ is chosen using the approach introduced in Sect. 4.1.2 that guarantees the averaged performance level $\epsilon(\eta = 10) < 0.1$ under a uniform sampling distribution (see Example 4.3).

In the figures of simulation results, the target positions, the initial positions of agents, the past sampling positions of agents, and the current positions of agents are represented by white stars, yellow crosses, pink dots, and white circles with agent indices, respectively.

4.3.1 Centralized Sampling Scheme

Consider a situation where a central station has access to all measurements collected by agents. At each time, measurements sampled by agents are transmitted to the central station that uses the centralized navigation strategy and sends control commands back to individual agents.

Case 1: First, we consider a set of fixed target points, e.g., 6×6 grid points on $\mathcal{Q}$ at a fixed time $t = 10$. At each time step, the cost function J_c in (4.16), which is the average of prediction error variances at target points, is minimized due to the proposed centralized navigation strategy in Sect. 4.2.1. As a benchmark strategy, we consider a random sampling scheme in which a group of 9 agents takes observations at randomly selected positions within the surveillance region $\mathcal{Q}$.

In Fig. 4.5a, the blue circles represent the average of prediction error variances over target points achieved by the centralized scheme, and the red squares indicate the

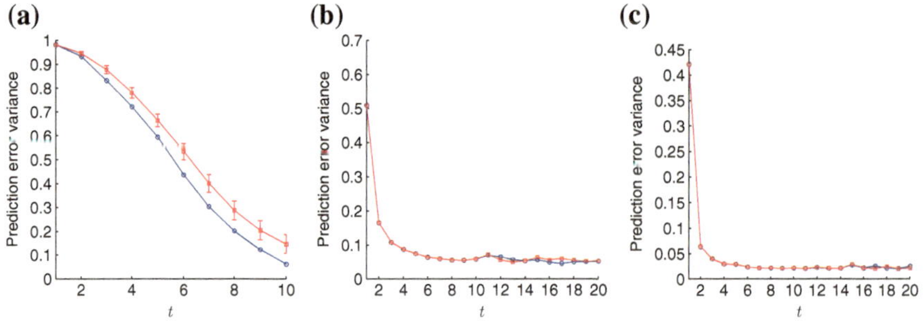

Fig. 4.5 Average of prediction error variances over target points (in *blue circles*) achieved by the centralized sampling scheme using all collective observations for **a** Case 1, **b** Case 2, and **c** Case 3. In (**a**), the target points are fixed at time $t = 10$, and the counterpart achieved by the benchmark random sampling strategy is shown in *red squares* with error-bars. In (**b**) and (**c**), the target points are at $t + 1$ and change over time. The counterpart achieved by using truncated observations are shown in *red squares*

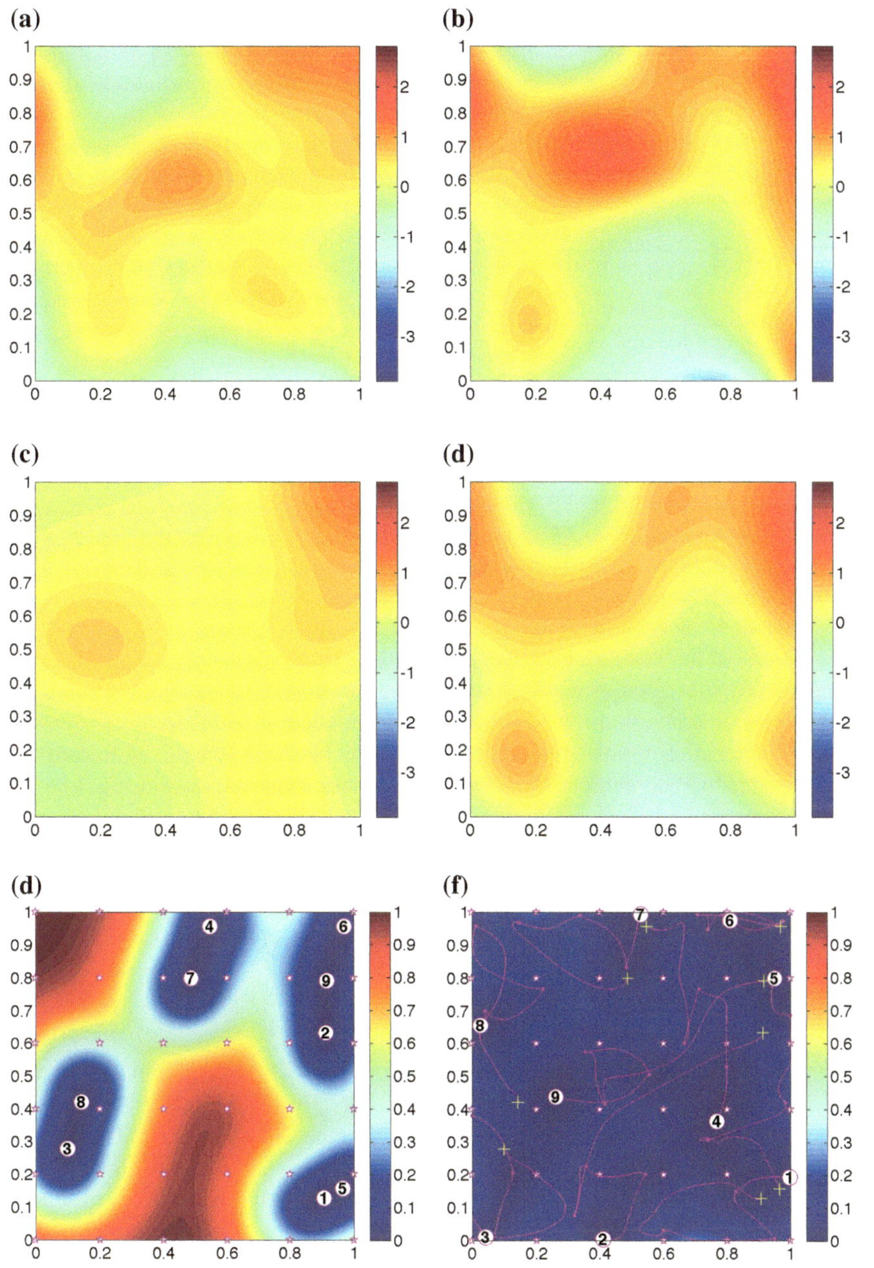

Fig. 4.6 Simulation results at $t = 1$ and $t = 5$ obtained by the centralized sampling scheme for Case 2. **a** True field at $t = 1$. **b** True field at $t = 5$. **c** Predicted field at $t = 1$. **d** Predicted field at $t = 5$. **e** Prediction error variance at $t = 1$. **f** Prediction error variance at $t = 5$.

average of prediction error variances over target points achieved by the benchmark strategy. Clearly, the proposed scheme produces lower averaged prediction error variances at target points as time increases, which demonstrates the usefulness of our scheme.

Case 2: Next, we consider the same 6×6 grid points on $\mathcal{Q}$ as in Case 1. However, at time t, we are now interested in the prediction at the next sampling time $t+1$. At each time step, the cost function J_c is minimized. Figure 4.5b shows the average of prediction error variances over target points achieved by the centralized scheme with truncation (in red squares) and without truncation (in blue circles). With truncated observations, i.e., with only observations obtained from latest $\eta = 10$ time steps, we are able to maintain the same level of the averaged prediction error variances (around 0.05 in Fig. 4.5b).

Figure 4.6a, c and e show the true field, the predicted field, and the prediction error variance at time $t = 1$, respectively. To see the improvement, the counterpart of the simulation results at time $t = 5$ are shown in Fig. 4.6b, d and f. At time $t = 1$, agents have little information about the field and hence the prediction is far away from the true field, which produces a large prediction error variance. As time increases, the prediction becomes close to the true field and the prediction error variances are reduced due to the proposed navigation strategy.

Case 3: Now, we consider another case in which 36 target points (plotted in Fig. 4.7 as white stars) are evenly distributed on three concentric circles to form a ring shaped subregion of interest. As in Case 2, we are interested in the prediction at the next time iteration $t+1$. The average of prediction error variances over these target points at each time step achieved by the centralized scheme with truncation (in red squares) and without truncation (in blue circles) are shown in Fig. 4.5c. The prediction error variances at time $t = 1$ and $t = 5$ are shown in Fig. 4.7a and b, respectively. It is shown that agents are dynamically covering the ring shaped region to minimize the average of prediction error variances over the target points.

4.3.2 Distributed Sampling Scheme

Consider a situation in which the sensor network has a limited communication range R, i.e., $\mathcal{N}_i(t) := \left\{ j \in \mathcal{I} \mid \left\| \mathbf{q}_i(t) - \mathbf{q}_j(t) \right\| < R, j \neq i \right\}$. At each time $t \in \mathbb{Z}_{>0}$, agent i collects measurements from itself and its neighbors $\mathcal{N}_i(t)$ and makes prediction in a distributed fashion. The distributed strategy is used to navigate itself to move to the next sampling position. To be comparable with the centralized scheme, the same target points as in Case 2 of Sect. 4.3.1 are considered.

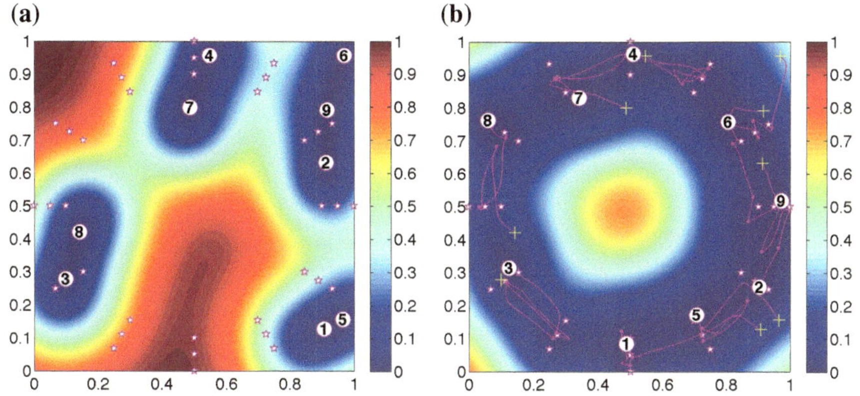

Fig. 4.7 Simulation results obtained by the centralized sampling scheme for Case 3. The trajectories of agents are shown in *solid lines* **a** Prediction error variance at $t = 1$. **b** Prediction error variance at $t = 5$

Figure 4.8 shows that the cost function, which is an upper-bound of the averaged prediction error variance over target points and agents, deceases smoothly from time $t = 1$ to $t = 2$ by the gradient descent algorithm with a communication range $R = 0.4$. Significant decreases occur whenever one of the agent gains a neighbor. Notice that the discontinuity of minimizing J in (4.20) caused by gaining or losing neighbors is eliminated due to the construction of J_d in (4.23). Hence, the proposed distributed algorithm is robust to gaining or losing neighbors.

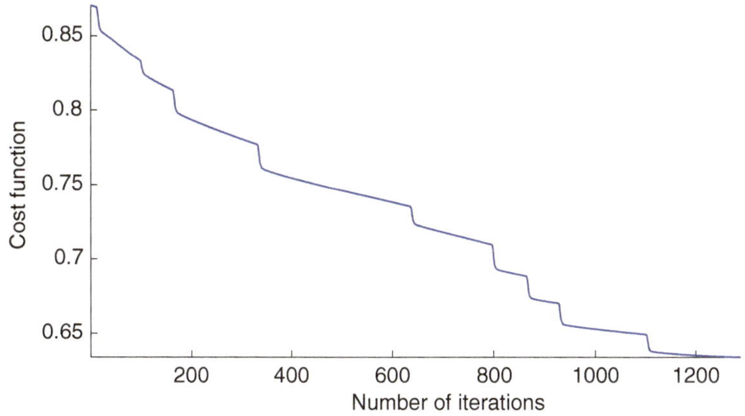

Fig. 4.8 Cost function $J_d(\tilde{\mathbf{q}})$ from $t = 1$ to $t = 2$ with a communication range $R = 0.4$

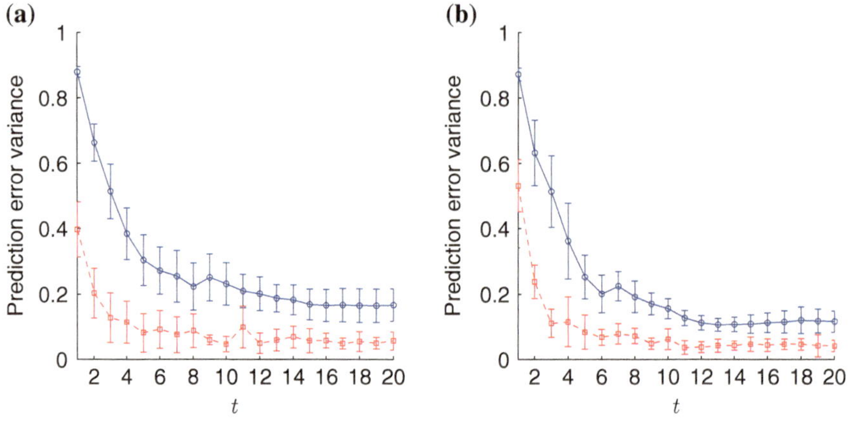

Fig. 4.9 Average of prediction error variances over all target points and agents achieved by the distributed sampling scheme with a communication range **a** $R = 0.3$, and **b** $R = 0.4$. The average of prediction error variances over all target points and agents are shown in blue circles. The average of prediction error variance over local target points and agents are shown in red squares. The error-bars indicate the standard deviation among agents

The following study shows the effect of different communication range. Intuitively, the larger the communication range is, the more information can be obtained by the agent and hence the better prediction can be made. Figures 4.9a and b show the average of prediction error variances over all target points and agents in blue circles with error-bars indicating the standard deviation among agents for the case $R = 0.3$ and $R = 0.4$, respectively. In both cases, $d_0 = 0.1$ in (4.22) was used. The average of prediction error variances is minimized quickly to a certain level. It can be seen that the level of achieved averaged prediction error variance with $R = 0.4$ is lower than the counterpart with $R = 0.3$.

Now, assume that each agent only predict the field at target points within radius R (local target points). The average of prediction error variances, over only local target points and agents, are also plotted in Fig. 4.9 in red squares with the standard deviation among agents. As can be seen, the prediction error variances at local target points (the red squares) are significantly lower than those for all target points (the blue circles).

Figure 4.10 shows the prediction error variances obtained by agent 1 along with the edges of the communication network for different communication range R and different time step t. In Fig. 4.10, the target positions, the initial positions, and the current positions are represented by white stars, yellow crosses, and white circles, respectively. Surprisingly, the agents under the distributed navigation algorithm produce an emergent, swarm-like behavior to maintain communication connectivity

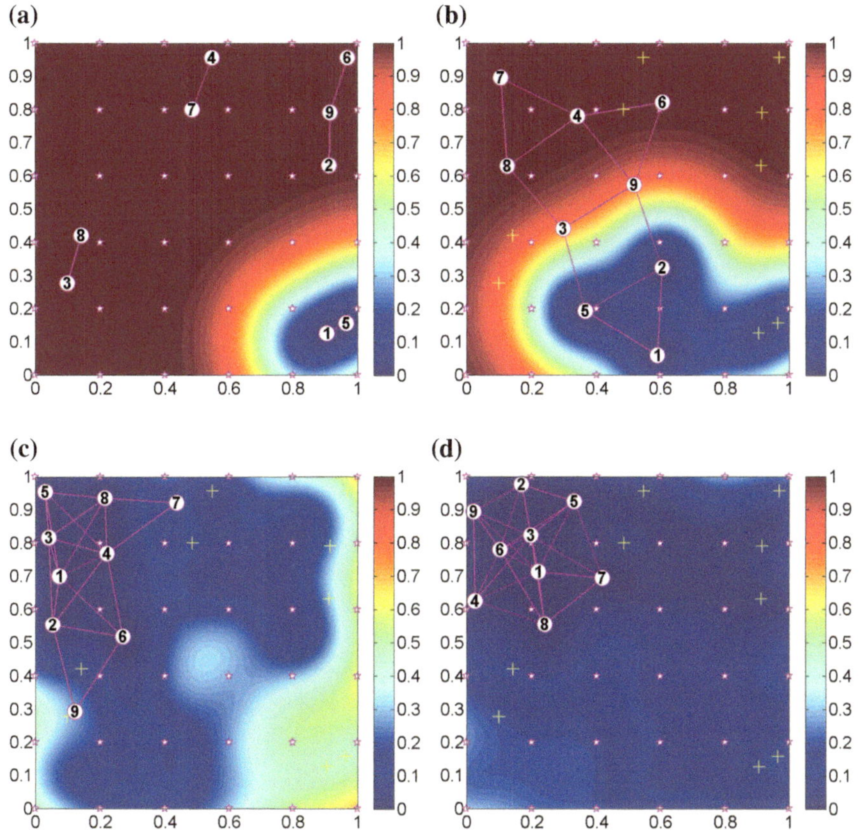

Fig. 4.10 Simulation results obtained by the distributed sampling scheme with different communication ranges. The edges of the graph are shown in *solid lines*. **a** $R = 0:3, t = 1$. **b** $R = 0:3, t = 2$. **c** $R = 0:3, t = 5$. **d** $R = 0:3, t = 20$. **e** $R = 0:4, t = 1$. **f** $R = 0:4, t = 2$. **g** $R = 0:4, t = 5$. **h** $R = 0:4, t = 20$

among local neighbors. Notice that this collective behavior emerged naturally and was not generated by the flocking or swarming algorithm as in [42].

This interesting simulation study (Fig. 4.10) shows that agents won't get too close each other since the average of prediction error variances at target points can be reduced by spreading over and covering the target points that need to be sampled. However, agents won't move too far away each other since the average of prediction error variances can be reduced by collecting measurements from a larger population of neighbors. This trade-off is controlled by the communication range. With the

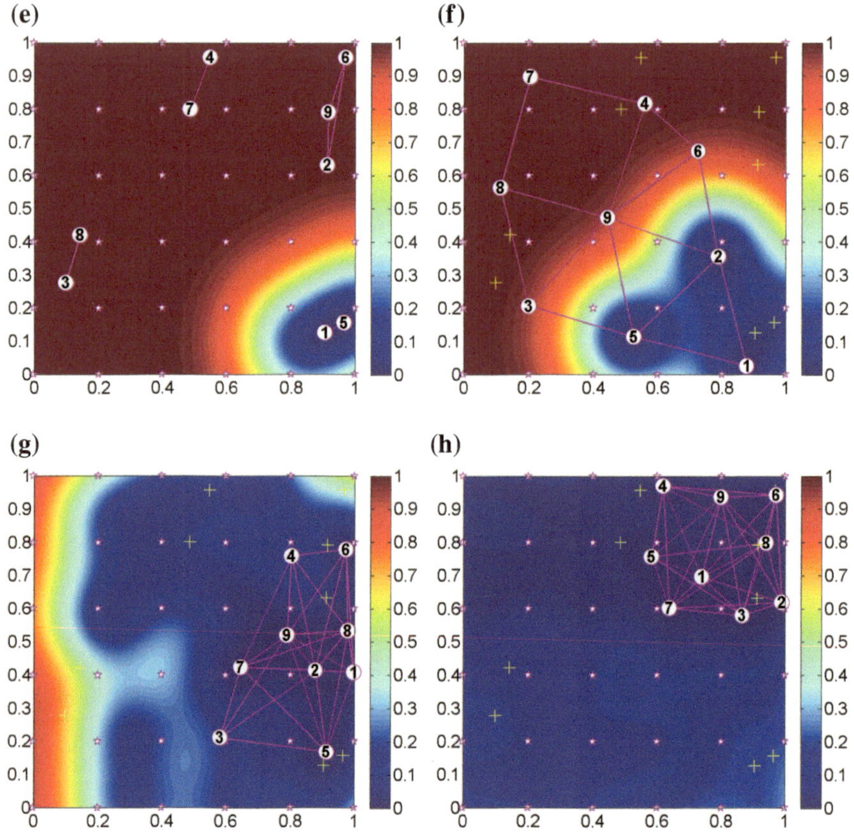

Fig. 4.10 (continued)

intertwined dynamics of agents over the proximity graph, as shown in Fig. 4.10, mobile sensing agents are coordinated in each time iteration in order to dynamically cover the target positions for better collective prediction capability.

Chapter 5
Fully Bayesian Approach

In Chap. 4, we analyzed the conditions under which near-optimal prediction can be achieved using only truncated observations. This motivates the usage of sparse Gaussian process proposed in [108]. However, they both assumed the covariance function is known a priori. In this chapter, we relax this stringent assumption. Unknown parameters in the covariance function can be estimated by a point estimator based on maximum likelihood (ML) or maximum *a posterior* (MAP). Such ML or MAP estimates may be regarded as the true parameters and then used in the prediction as in Chap. 3. However, the point estimate itself needs to be identified using sufficient amount of measurements and most importantly, the whole procedure fails to incorporate the uncertainty in the point estimate into the prediction.

Motivated by the aforementioned issues, a fully Bayesian framework is adopted in this chapter as it offers several advantages when inferring parameters and processes from highly complex models. The Bayesian approach requires prior distributions to be elicited for model parameters that are of interest. Although difficult initially, this forces the practitioner to reflect on all sources and extent of uncertainties from scientific knowledge and past experience. In highly complex models, this pre-analysis of uncertainties eventually yields better inference for the parameters of interest, and even more so when data contains limited information. Once the priors are elicited, the Bayesian framework is flexible and effective in incorporating all uncertainties as well as information (limited or otherwise from data) into a single entity, namely, the posterior. There is, thus, no need to pre-process or account for each source of uncertainty separately. The Bayesian framework seamlessly describes their joint influence and contributions through the posterior distribution—which is also, conveniently, the only entity to focus on for inference. The fully Bayesian approach further advocates priors for all unknown entities in the model; these entities can either be nuisance parameters (such as the scale of measurement error from each mobile sensor) or (hyper) parameters that govern the prior distributions (such as the

Y. Xu et al., *Bayesian Prediction and Adaptive Sampling Algorithms for Mobile Sensor Networks*, SpringerBriefs in Control, Automation and Robotics, DOI 10.1007/978-3-319-21921-9_5

extent of spatial variability of the scalar field). The fully Bayesian approach thus allows additional sources and extent of uncertainties to be integrated into the inferential framework, with the posterior distribution effectively capturing all aspects of uncertainties involved. Subsequently, the practitioner needs only to focus on different components of the posterior to obtain inference separately for the parameters of interest, nuisance parameters and hyperparameters. The fully Bayesian approach also allows data to select the most appropriate values for nuisance parameters and hyperparameters automatically, and achieve optimal inference and prediction for the scalar field.

The advantage of a fully Bayesian approach, which will be adopted in this chapter, is that the uncertainty in the model parameters are incorporated in the prediction [84]. In [85], Gaudard et al. presented a Bayesian method that uses importance sampling for analyzing spatial data sampled from a Gaussian random field whose covariance function was unknown. However, the assumptions made in [85], such as noiseless observations and time-invariance of the field, limit the applicability of the approach on mobile sensors in practice. The computational complexity of a fully Bayesian prediction algorithm has been the main hurdle for applications in resource-constrained robots. In [46], an iterative prediction algorithm without resorting to Markov Chain Monte Carlo (MCMC) methods has been developed based on analytical closed form solutions from results in [85], by assuming that the covariance function of the spatiotemporal Gaussian random field is known up to a constant. Our work builds on such Bayesian approaches used in [46, 85] and explores new ways to synthesize practical algorithms for mobile sensor networks under more relaxed conditions.

In Sect. 5.1, we provide fully Bayesian approaches for spatiotemporal Gaussian process regression under more practical conditions such as measurement noise and the unknown covariance function. In Sect. 5.2, using discrete prior probabilities and compactly supported kernels, we provide a way to design sequential Bayesian prediction algorithms in which the exact predictive distributions can be computed in constant time (i.e., $O(1)$) as the number of observations increases. In particular, a centralized sequential Bayesian prediction algorithm is developed in Sect. 5.2.1, and its distributed implementation among sensor groups is provided for a special case in Sect. 5.2.2. An adaptive sampling strategy for mobile sensors, utilizing the maximum *a posteriori* (MAP) estimation of the parameters, is proposed to minimize the prediction error variances in Sect. 5.2.3. In Sect. 5.3, the proposed sequential Bayesian prediction algorithms and the adaptive sampling strategy are tested under practical conditions for spatiotemporal Gaussian processes.

5.1 Fully Bayesian Prediction Approach

In this chapter, we consider a spatiotemporal Gaussian process denoted by

$$z(\mathbf{x}) \sim \mathcal{GP}\left(\mu(\mathbf{x}), \sigma_f^2 C(\mathbf{x}, \mathbf{x}'; \theta)\right),$$

where $z(\mathbf{x}) \in \mathbb{R}$ and $\mathbf{x} := (\mathbf{s}^T, t)^T \in \mathcal{Q} \times \mathbb{Z}_{>0}$ contains the sampling location $\mathbf{s} \in \mathcal{Q} \subset \mathbb{R}^2$ and the sampling time $t \in \mathbb{Z}_{>0}$. The mean function is assumed to be

$$\mu(\mathbf{x}) = f(\mathbf{x})^T \beta,$$

where $f(\mathbf{x}) := (f_1(\mathbf{x}), \ldots, f_p(\mathbf{x}))^T \in \mathbb{R}^p$ is a known regression function, and $\beta \in \mathbb{R}^p$ is an unknown vector of regression coefficients. The correlation between $z(\mathbf{x})$ and $z(\mathbf{x}')$ is taken as

$$C(\mathbf{x}, \mathbf{x}'; \theta) = C_s\left(\frac{\|\mathbf{s} - \mathbf{s}'\|}{\sigma_s}\right) C_t\left(\frac{|t - t'|}{\sigma_t}\right), \tag{5.1}$$

which is governed by spatial and temporal distance functions $C_s(\cdot)$ and $C_t(\cdot)$. We assume that $C_s(\cdot)$ and $C_t(\cdot)$ are decreasing kernel functions over space and time, respectively, so that the correlation between two inputs decreases as the distance between spatial locations (respectively, time indices) increases. The decreasing rate depends on the spatial bandwidth σ_s (respectively, the time bandwidth σ_t) for given fixed time indices (respectively, spatial locations). The signal variance σ_f^2 gives the overall vertical scale relative to the mean of the Gaussian process in the output space. We define $\boldsymbol{\theta} := (\sigma_s, \sigma_t)^T \in \mathbb{R}^2$ for notational simplicity.

Given the collection of noise-corrupted observations from mobile sensing agents up to time t, we want to predict $z(\mathbf{s}_*, t_*)$ at a prespecified location $\mathbf{s}_* \in \mathcal{S} \subset \mathcal{Q}$ and current (or future) time t_*. To do this, suppose we have a collection of n observations $\mathcal{D} = \left\{(\mathbf{x}^{(i)}, y^{(i)}) \mid i = 1, \ldots, n\right\}$ from N mobile sensing agents up to time t. Here $\mathbf{x}^{(i)}$ denotes the ith input vector of dimension 3 (i.e., the sampling position and time of the ith observation) and $y^{(i)}$ denotes the ith noise-corrupted measurement. If all observations are considered, we have $n = Nt$. Notice that the number of observations n grows with the time t. For notational simplicity, let $\mathbf{y} := (y^{(1)}, \ldots, y^{(n)})^T \in \mathbb{R}^n$ denote the collection of noise-corrupted observations. Based on the spatiotemporal Gaussian process, the distribution of the observations given the parameters β, σ_f^2, and $\boldsymbol{\theta}$ is Gaussian, i.e.,

$$\mathbf{y}|\beta, \sigma_f^2, \boldsymbol{\theta} \sim \mathcal{N}(\mathbf{F}\beta, \sigma_f^2 \mathbf{C})$$

with $\mathbf{F}$ and $\mathbf{C}$ defined as

$$\begin{aligned} \mathbf{F} &:= \left(f(\mathbf{x}^{(1)}), \ldots, f(\mathbf{x}^{(n)})\right)^T \in \mathbb{R}^{n \times p}, \\ \mathbf{C} &:= \mathrm{Corr}(\mathbf{y}, \mathbf{y}|\theta) = \left[C(\mathbf{x}^{(i)}, \mathbf{x}^{(j)}; \theta) + \frac{1}{\gamma}\delta_{ij}\right] \in \mathbb{R}^{n \times n}, \end{aligned} \tag{5.2}$$

where δ_{ij} is the Kronecker delta which equals to one when $i = j$, and zero otherwise.

5.1.1 Prior Selection

To infer the unknown parameters $\boldsymbol{\beta}, \sigma_f^2$, and $\boldsymbol{\theta}$ in a Bayesian framework, the collection of them is considered to be a random vector with a prior distribution reflecting the *a priori* belief of uncertainty for them. In this chapter, we use the prior distribution given by

$$\pi(\boldsymbol{\beta}, \sigma_f^2, \boldsymbol{\theta}) = \pi(\boldsymbol{\beta}|\sigma_f^2)\pi(\sigma_f^2)\pi(\boldsymbol{\theta}), \tag{5.3}$$

where

$$\boldsymbol{\beta}|\sigma_f^2 \sim \mathbb{N}(\boldsymbol{\beta}_0, \sigma_f^2 \mathbf{T}).$$

The prior for $\pi(\sigma_f^2)$ is taken to be the inverse gamma distribution, chosen to guarantee positiveness of σ_f^2 and a closed-form expression for the posterior distribution of σ_f^2 for computational ease of the proposed algorithms. To cope with the case where no prior knowledge on $\boldsymbol{\beta}$ is available, which is often the case in practice, we propose to use a noninformative prior. In particular, we take $\boldsymbol{\beta}_0 = \mathbf{0}, \mathbf{T} = \alpha \mathbf{I}$, and subsequently, let $\alpha \to \infty$. Any proper prior $\pi(\boldsymbol{\theta})$ that correctly reflects the priori knowledge of $\boldsymbol{\theta}$ can be used.

5.1.2 MCMC-Based Approach

According to the Bayes rule, the posterior distribution of $\boldsymbol{\beta}$, σ_f^2, and $\boldsymbol{\theta}$ is given by

$$\pi(\boldsymbol{\beta}, \sigma_f^2, \boldsymbol{\theta}|y) = \frac{\pi(y|\boldsymbol{\beta}, \sigma_f^2, \boldsymbol{\theta})\pi(\boldsymbol{\beta}, \sigma_f^2, \boldsymbol{\theta})}{\int\int\int \pi(y|\boldsymbol{\beta}, \sigma_f^2, \boldsymbol{\theta})\pi(\boldsymbol{\beta}, \sigma_f^2, \boldsymbol{\theta}) d\boldsymbol{\beta} d\sigma_f^2 d\boldsymbol{\theta}}. \tag{5.4}$$

When a proper prior is used, the posterior distribution can be written as

$$\pi(\boldsymbol{\beta}, \sigma_f^2, \boldsymbol{\theta}|y) \propto \pi(y|\boldsymbol{\beta}, \sigma_f^2, \boldsymbol{\theta})\pi(\boldsymbol{\beta}, \sigma_f^2, \boldsymbol{\theta}).$$

The inference on $\boldsymbol{\beta}$, σ_f^2, and $\boldsymbol{\theta}$ can be carried out by sampling from the posterior distribution in (5.4) via the Gibbs sampler. Table 5.1 gives the steps based on the following proposition.

Proposition 5.1 *For a prior distribution given in* (5.3) *with the noninformative prior on* $\boldsymbol{\beta}$, *the conditional posteriors are given by*

1. $\boldsymbol{\beta}|\sigma_f^2, \boldsymbol{\theta}, \mathbf{y} \sim \mathbb{N}\left(\hat{\boldsymbol{\beta}}, \sigma_f^2 \boldsymbol{\Sigma}_{\hat{\boldsymbol{\beta}}}\right)$, *where*

$$\boldsymbol{\Sigma}_{\hat{\boldsymbol{\beta}}} = (\mathbf{F}^T \mathbf{C}^{-1} \mathbf{F})^{-1},$$
$$\hat{\boldsymbol{\beta}} = \boldsymbol{\Sigma}_{\hat{\boldsymbol{\beta}}}(\mathbf{F}^T \mathbf{C}^{-1} \mathbf{y}).$$

Table 5.1 Gibbs sampler

Input: initial samples $\boldsymbol{\beta}^{(1)}$, $\sigma_f^{2\,(1)}$, and $\boldsymbol{\theta}^{(1)}$
Output: samples $\left\{\boldsymbol{\beta}^{(i)}, \sigma_f^{2\,(i)}, \boldsymbol{\theta}^{(i)}\right\}_{i=1}^{m}$ from joint distribution $\pi(\boldsymbol{\beta}, \sigma_f^2, \boldsymbol{\theta}\vert\mathbf{y})$
1: initialize $\boldsymbol{\beta}^{(1)}, \sigma_f^{2\,(1)}, \boldsymbol{\theta}^{(1)}$
2: **for** $i = 1$ to m **do**
3: sample $\boldsymbol{\beta}^{(i+1)}$ from $\pi(\boldsymbol{\beta}\vert\sigma_f^{2\,(i)}, \boldsymbol{\theta}^{(i)}, \mathbf{y})$
4: sample $\sigma_f^{2\,(i+1)}$ from $\pi(\sigma_f^2\vert\boldsymbol{\beta}^{(i+1)}, \boldsymbol{\theta}^{(i)}, \mathbf{y})$
5: sample $\boldsymbol{\theta}^{(i+1)}$ from $\pi(\boldsymbol{\theta}\vert\boldsymbol{\beta}^{(i+1)}, \sigma_f^{2\,(i+1)}, \mathbf{y})$
6: **end for**

2. $\sigma_f^2|\beta, \boldsymbol{\theta}, \mathbf{y} \sim IG\left(\bar{a}, \bar{b}\right)$, *where*

$$\bar{a} = a + \frac{n+p}{2},$$
$$\bar{b} = b + \frac{1}{2}(\mathbf{y} - \mathbf{F}\beta)^T \mathbf{C}^{-1}(\mathbf{y} - \mathbf{F}\beta).$$

3.
$$\pi(\boldsymbol{\theta}|\beta, \sigma_f^2, \mathbf{y}) \propto \det(\mathbf{C})^{-1/2} \exp\left(-\frac{(\mathbf{y} - \mathbf{F}\beta)^T \mathbf{C}^{-1}(\mathbf{y} - \mathbf{F}\beta)}{2\sigma_f^2}\right)\pi(\boldsymbol{\theta}).$$

Proof Since the noninformative prior is chosen, the posterior distribution shall be computed with $\mathbf{T} = \alpha\mathbf{I}$ and then let $\alpha \to \infty$.
(i) For given σ_f^2, $\boldsymbol{\theta}$, and $\mathbf{y}$, we have

$$\pi(\beta|\sigma_f^2, \boldsymbol{\theta}, \mathbf{y}) = \lim_{\alpha\to\infty} \frac{\pi(y|\beta, \sigma_f^2, \boldsymbol{\theta})\pi(\beta|\sigma_f^2)}{\int \pi(y|\beta, \sigma_f^2, \boldsymbol{\theta})\pi(\beta|\sigma_f^2)d\beta}.$$

Let

$$\begin{aligned}
\mathrm{num}_1 &= \pi(\mathbf{y}|\beta, \sigma_f^2, \boldsymbol{\theta})\pi(\beta|\sigma_f^2) \\
&= \frac{\exp\left\{-\frac{1}{2\sigma_f^2}(\mathbf{y} - \mathbf{F}\beta)^T C^{-1}(\mathbf{y} - \mathbf{F}\beta)\right\}}{(2\pi\sigma_f^2)^{n/2}\det(\mathbf{C})^{1/2}} \frac{\exp\left\{-\frac{1}{2\sigma_f^2}\beta^T \mathbf{T}^{-1}\beta\right\}}{(2\pi\sigma_f^2)^{p/2}\det(\mathbf{T})^{1/2}} \\
&= \frac{\exp\left\{-\frac{1}{2\sigma_f^2} RSS\right\}}{(2\pi\sigma_f^2)^{(n+p)/2}\det(\mathbf{C})^{1/2}\det(\mathbf{T})^{1/2}} \\
&\times \exp\left\{-\frac{1}{2\sigma_f^2}(\beta - \hat{\beta})^T(\mathbf{F}^T\mathbf{C}^{-1}\mathbf{F} + \mathbf{T}^{-1})(\beta - \hat{\beta})\right\},
\end{aligned}$$

and

$$
\begin{aligned}
\texttt{den}_1 &= \frac{\exp\left\{-\frac{1}{2\sigma_f^2} RSS\right\}}{(2\pi\sigma_f^2)^{(n+p)/2}\det(\mathbf{C})^{1/2}\det(\mathbf{T})^{1/2}} \\
&\quad\times \int \exp\left\{-\frac{1}{2\sigma_f^2}(\beta-\hat{\beta})^T(\mathbf{F}^T\mathbf{C}^{-1}\mathbf{F}+\mathbf{T}^{-1})(\beta-\hat{\beta})\right\} d\beta \\
&= \frac{\exp\left\{-\frac{1}{2\sigma_f^2} RSS\right\}}{(2\pi\sigma_f^2)^{(n+p)/2}\det(\mathbf{C})^{1/2}\det(\mathbf{T})^{1/2}} (2\pi\sigma_f^2)^{p/2}\det(\mathbf{F}^T\mathbf{C}^{-1}\mathbf{F}+\mathbf{T}^{-1})^{-1/2}
\end{aligned}
$$

where

$$
\begin{aligned}
RSS &= \mathbf{y}^T\left(\mathbf{C}^{-1}-\mathbf{C}^{-1}\mathbf{F}(\mathbf{F}^T\mathbf{C}^{-1}\mathbf{F}+\mathbf{T}^{-1})^{-1}\mathbf{F}^T\mathbf{C}^{-1}\right)\mathbf{y} \\
&= \mathbf{y}^T(\mathbf{C}+\mathbf{F}\mathbf{T}\mathbf{F}^T)^{-1}\mathbf{y}.
\end{aligned}
$$

Then we have

$$
\begin{aligned}
\pi(\beta|\sigma_f^2, \boldsymbol{\theta}, y) &= \lim_{\alpha\to\infty}\frac{\texttt{num}_1}{\texttt{den}_1} \\
&= \lim_{\alpha\to\infty}\frac{\exp\left\{-\frac{1}{2\sigma_f^2}(\beta-\hat{\beta})^T(\mathbf{F}^T\mathbf{C}^{-1}\mathbf{F}+\mathbf{T}^{-1})(\beta-\hat{\beta})\right\}}{(2\pi\sigma_f^2)^{p/2}\det(\mathbf{F}^T\mathbf{C}^{-1}\mathbf{F}+\mathbf{T}^{-1})^{-1/2}} \\
&= \frac{\exp\left\{-\frac{1}{2\sigma_f^2}(\beta-\hat{\beta})^T\boldsymbol{\Sigma}_{\hat{\beta}}^{-1}(\beta-\hat{\beta})\right\}}{(2\pi\sigma_f^2)^{p/2}\det(\boldsymbol{\Sigma}_{\hat{\beta}})^{1/2}}.
\end{aligned}
$$

Therefore, we have $\beta|\sigma_f^2, \boldsymbol{\theta}, \mathbf{y} \sim \mathbb{N}(\hat{\beta}, \sigma_f^2\boldsymbol{\Sigma}_{\hat{\beta}})$.
(ii) For given $\boldsymbol{\beta}$, $\boldsymbol{\theta}$, and $\mathbf{y}$, we have

$$
\begin{aligned}
\pi(\sigma_f^2|\boldsymbol{\beta}, \boldsymbol{\theta}, \mathbf{y}) &= \lim_{\alpha\to\infty}\frac{\pi(\mathbf{y}|\boldsymbol{\beta}, \sigma_f^2, \boldsymbol{\theta})\pi(\sigma_f^2|\boldsymbol{\beta})}{\int\pi(\mathbf{y}|\boldsymbol{\beta}, \sigma_f^2, \boldsymbol{\theta})\pi(\sigma_f^2|\boldsymbol{\beta})d\sigma_f^2} \\
&= \lim_{\alpha\to\infty}\frac{\pi(\mathbf{y}|\boldsymbol{\beta}, \sigma_f^2, \boldsymbol{\theta})\pi(\boldsymbol{\beta}|\sigma_f^2)\pi(\sigma_f^2)}{\int\pi(\mathbf{y}|\boldsymbol{\beta}, \sigma_f^2, \boldsymbol{\theta})\pi(\boldsymbol{\beta}|\sigma_f^2)\pi(\sigma_f^2)d\sigma_f^2}.
\end{aligned}
$$

Let

$$
\begin{aligned}
\mathtt{num}_2 &= \pi(\mathbf{y}|\boldsymbol{\beta}, \sigma_f^2, \boldsymbol{\theta})\pi(\boldsymbol{\beta}|\sigma_f^2)\pi(\sigma_f^2) \\
&= \frac{\exp\left\{-\frac{1}{2\sigma_f^2}(\mathbf{y}-\mathbf{F}\beta)^T\mathbf{C}^{-1}(\mathbf{y}-\mathbf{F}\beta)\right\}}{(2\pi\sigma_f^2)^{n/2}\det(\mathbf{C})^{1/2}} \\
&\times \frac{\exp\left\{-\frac{1}{2\sigma_f^2}\beta^T\mathbf{T}^{-1}\beta\right\}}{(2\pi\sigma_f^2)^{p/2}\det(\mathbf{T})^{1/2}} \frac{b^a\exp\left\{-\frac{b}{\sigma_f^2}\right\}}{\Gamma(a)(\sigma_f^2)^{a+1}} \\
&= \frac{b^a}{\Gamma(a)(2\pi)^{\bar{a}+1}\det(\mathbf{C})^{1/2}\det(\mathbf{T})^{1/2}} \frac{1}{(\sigma_f^2)^{\bar{a}+1}} \exp\left\{-\frac{\bar{b}+\frac{1}{2}\boldsymbol{\beta}^T\mathbf{T}^{-1}\boldsymbol{\beta}}{\sigma_f^2}\right\},
\end{aligned}
$$

and

$$
\begin{aligned}
\mathtt{den}_2 &= \frac{b^a}{\Gamma(a)(2\pi)^{\bar{a}+1}\det(\mathbf{C})^{1/2}\det(\mathbf{T})^{1/2}} \\
&\times \int \frac{1}{(\sigma_f^2)^{\bar{a}+1}} \exp\left\{-\frac{\bar{b}+\frac{1}{2}\boldsymbol{\beta}^T\mathbf{T}^{-1}\boldsymbol{\beta}}{\sigma_f^2}\right\} d\sigma_f^2 \\
&= \frac{b^a}{\Gamma(a)(2\pi)^{\bar{a}+1}\det(\mathbf{C})^{1/2}\det(\mathbf{T})^{1/2}} \Gamma(\bar{a})\bar{b}^{-\bar{a}}.
\end{aligned}
$$

Then we have

$$
\begin{aligned}
\pi(\sigma_f^2|\boldsymbol{\beta}, \boldsymbol{\theta}, \mathbf{y}) &= \lim_{\alpha\to\infty} \frac{\mathtt{num}_2}{\mathtt{den}_2} \\
&= \lim_{\alpha\to\infty} \frac{\bar{b}^{\bar{a}}}{\Gamma(\bar{a})(\sigma_f^2)^{\bar{a}+1}} \exp\left\{-\frac{\bar{b}+\frac{1}{2}\boldsymbol{\beta}^T\mathbf{T}^{-1}\boldsymbol{\beta}}{2\sigma_f^2}\right\} \\
&= \frac{\bar{b}^{\bar{a}}}{\Gamma(\bar{a})(\sigma_f^2)^{\bar{a}+1}} \exp\left\{-\frac{\bar{b}}{2\sigma_f^2}\right\}.
\end{aligned}
$$

Therefore, we have $\sigma_f^2|\boldsymbol{\beta}, \boldsymbol{\theta}, \mathbf{y} \sim IG\left(\bar{a}, \bar{b}\right)$.
(iii) For given $\boldsymbol{\beta}$, σ_f^2, and $\mathbf{y}$, we have

$$
\begin{aligned}
\pi(\boldsymbol{\theta}|\boldsymbol{\beta}, \sigma_f^2, \mathbf{y}) &= \lim_{\alpha\to\infty} \frac{\pi(\mathbf{y}|\boldsymbol{\beta}, \sigma_f^2, \boldsymbol{\theta})\pi(\boldsymbol{\theta})}{\int \pi(\mathbf{y}|\boldsymbol{\beta}, \sigma_f^2, \boldsymbol{\theta})\pi(\boldsymbol{\theta})d\boldsymbol{\theta}} \\
&\propto \det(\mathbf{C})^{-1/2} \exp\left(-\frac{(\mathbf{y}-\mathbf{F}\beta)^T\mathbf{C}^{-1}(\mathbf{y}-\mathbf{F}\beta)}{2\sigma_f^2}\right)\pi(\boldsymbol{\theta}).
\end{aligned}
$$

The posterior predictive distribution of $z_* := z(\mathbf{s}_*, t_*)$ at location $\mathbf{s}_*$ and time t_* can be obtained by

$$\pi(z_*|\mathbf{y}) = \iiint \pi(z_*|\mathbf{y}, \boldsymbol{\beta}, \sigma_f^2, \boldsymbol{\theta})\pi(\boldsymbol{\beta}, \sigma_f^2, \boldsymbol{\theta}|\mathbf{y})d\boldsymbol{\beta}d\sigma_f^2 d\boldsymbol{\theta}, \tag{5.5}$$

where in (5.5), the conditional distribution $\pi(z_*|\boldsymbol{\beta}, \sigma_f^2, \boldsymbol{\theta}, \mathbf{y})$, is integrated with respect to the posterior of $\boldsymbol{\beta}$, σ_f^2, and $\boldsymbol{\theta}$ given observations $\mathbf{y}$. The conditional distribution of z_* is Gaussian, i.e.,

$$z_*|\boldsymbol{\beta}, \sigma_f^2, \boldsymbol{\theta}, \mathbf{y} \sim \mathbb{N}(\mu_{z_*|\boldsymbol{\beta},\sigma_f^2,\boldsymbol{\theta},\mathbf{y}}, \sigma^2_{z_*|\boldsymbol{\beta},\sigma_f^2,\boldsymbol{\theta},\mathbf{y}}),$$

with

$$\begin{aligned}
\mu_{z_*|\boldsymbol{\beta},\sigma_f^2,\boldsymbol{\theta},\mathbf{y}} &= \mathbb{E}(z_*|\boldsymbol{\beta}, \sigma_f^2, \boldsymbol{\theta}, \mathbf{y}) = f(\mathbf{x}_*)^T\boldsymbol{\beta} + \mathbf{k}^T\mathbf{C}^{-1}(\mathbf{y} - \mathbf{F}\boldsymbol{\beta}),\\
\sigma^2_{z_*|\boldsymbol{\beta},\sigma_f^2,\boldsymbol{\theta},\mathbf{y}} &= \mathrm{Var}(z_*|\boldsymbol{\beta}, \sigma_f^2, \boldsymbol{\theta}, \mathbf{y}) = \sigma_f^2(1 - \mathbf{k}^T\mathbf{C}^{-1}\mathbf{k}),
\end{aligned}$$

where $\mathbf{k} := \mathrm{Corr}(\mathbf{y}, z_*|\boldsymbol{\theta}) = [C(\mathbf{x}^{(i)}, \mathbf{x}_*; \boldsymbol{\theta})] \in \mathbb{R}^n$. To obtain numerical values of $\pi(z_*|\mathbf{y})$, we draw m samples $\left\{\boldsymbol{\beta}^{(i)}, \sigma_f^{2\,(i)}, \boldsymbol{\theta}^{(i)}\right\}_{i=1}^m$ from the posterior distribution $\pi(\boldsymbol{\beta}, \sigma_f^2, \boldsymbol{\theta}|\mathbf{y})$ using the Gibbs sampler presented in Table 5.1, and then obtain the predictive distribution in (5.5) by

$$\pi(z_*|\mathbf{y}) \approx \frac{1}{m}\sum_{i=1}^m \pi(z_*|\mathbf{y}, \boldsymbol{\beta}^{(i)}, \sigma_f^{2\,(i)}, \boldsymbol{\theta}^{(i)}).$$

It follows that the predictive mean and variance can be obtained numerically by

$$\begin{aligned}
\mu_{z_*|\mathbf{y}} &= \mathbb{E}(z_*|\mathbf{y}) \approx \frac{1}{m}\sum_{i=1}^m \mu_{z_*|\boldsymbol{\beta}^{(i)},\sigma_f^{2\,(i)},\boldsymbol{\theta}^{(i)},\mathbf{y}},\\
\sigma^2_{z_*|\mathbf{y}} &= \mathrm{Var}(z_*|\mathbf{y}) \approx \frac{1}{m}\sum_{i=1}^m \sigma^2_{z_*|\boldsymbol{\beta}^{(i)},\sigma_f^{2\,(i)},\boldsymbol{\theta}^{(i)},\mathbf{y}}\\
&\quad + \frac{1}{m}\sum_{i=1}^m \left(\mu_{z_*|\boldsymbol{\beta}^{(i)},\sigma_f^{2\,(i)},\boldsymbol{\theta}^{(i)},\mathbf{y}} - \mu_{z_*|\mathbf{y}}\right)^2.
\end{aligned}$$

Remark 5.1 The Gibbs sampler presented in Table 5.1 may take long time to converge, which implies that the number of samples required could be quite large depending on the initial values. This convergence rate can be monitored from a trace plot (a plot of sampled values versus iterations for each variable in the chain). Moreover, since C is a complicated function of σ_s and σ_t, sampling from $\pi(\boldsymbol{\theta}|\boldsymbol{\beta}, \sigma_f^2, \mathbf{y})$ in Proposition 5.1 is difficult. An inverse cumulative distribution function (CDF) method

[109] needs to be used to generate samples, which requires griding on a continuous parameter space. Therefore, high computational power is needed to implement the MCMC-based approach.

In the next subsection, we present an alternative Bayesian approach which only requires drawing samples from the prior distribution $\pi(\boldsymbol{\theta})$ using a similar approach to one used in [85].

5.1.3 Importance Sampling Approach

The posterior predictive distribution of $z_* := z(\mathbf{s}_*, t_*)$ can be written as

$$\pi(z_*|\mathbf{y}) = \int \pi(z_*|\boldsymbol{\theta}, \mathbf{y})\pi(\boldsymbol{\theta}|\mathbf{y})d\boldsymbol{\theta}, \tag{5.6}$$

where

$$\pi(\boldsymbol{\theta}|\mathbf{y}) = \frac{\pi(\mathbf{y}|\boldsymbol{\theta})\pi(\boldsymbol{\theta})}{\int \pi(\mathbf{y}|\boldsymbol{\theta})\pi(\boldsymbol{\theta})d\boldsymbol{\theta}},$$

is the posterior distribution of $\boldsymbol{\theta}$, by integrating out analytically the parameters $\boldsymbol{\beta}$ and σ_f^2. We have the following proposition.

Proposition 5.2 *For a prior distribution given in* (5.3) *with the noninformative prior on* $\boldsymbol{\beta}$, *we have*

1. $\pi(\boldsymbol{\theta}|\mathbf{y}) \propto w(\boldsymbol{\theta}|\mathbf{y})\pi(\boldsymbol{\theta})$ *with*

$$\log w(\boldsymbol{\theta}|\mathbf{y}) = -\frac{1}{2}\log\det(\mathbf{C}) - \frac{1}{2}\log\det(\mathbf{F}^T\mathbf{C}^{-1}\mathbf{F}) - \tilde{a}\log\tilde{b}, \tag{5.7}$$

where

$$\tilde{a} = a + \frac{n}{2},$$
$$\tilde{b} = b + \frac{1}{2}\mathbf{y}^T\mathbf{C}^{-1}\mathbf{y} - \frac{1}{2}(\mathbf{F}^T\mathbf{C}^{-1}\mathbf{y})^T(\mathbf{F}^T\mathbf{C}^{-1}\mathbf{F})^{-1}(\mathbf{F}^T\mathbf{C}^{-1}\mathbf{y}).$$

2. $\pi(z_*|\boldsymbol{\theta}, \mathbf{y})$ *is a shifted student's t-distribution with location parameter* μ, *scale parameter* λ, *and* ν *degrees of freedom, i.e.,*

$$\pi(z_*|\boldsymbol{\theta}, \mathbf{y}) = \frac{\Gamma\left(\frac{\nu+1}{2}\right)}{\Gamma\left(\frac{\nu}{2}\right)}\left(\frac{\lambda}{\pi\nu}\right)^{\frac{1}{2}}\left(1 + \frac{\lambda(z_* - \mu)^2}{\nu}\right)^{-\frac{\nu+1}{2}}, \tag{5.8}$$

where $\nu = 2\tilde{a}$, *and*

$$\mu = \mathbf{k}^T\mathbf{C}^{-1}\mathbf{y} + (f(\mathbf{x}_*) - \mathbf{F}^T\mathbf{C}^{-1}\mathbf{k})^T(\mathbf{F}^T\mathbf{C}^{-1}\mathbf{F})^{-1}(\mathbf{F}^T\mathbf{C}^{-1}\mathbf{y}),$$

$$\lambda = \frac{\tilde{b}}{\tilde{a}}\big((1 - \mathbf{k}^T\mathbf{C}^{-1}\mathbf{k}) + (f(\mathbf{x}_*) - \mathbf{F}^T\mathbf{C}^{-1}\mathbf{k})^T(\mathbf{F}^T\mathbf{C}^{-1}\mathbf{F})^{-1}(f(\mathbf{x}_*) - \mathbf{F}^T\mathbf{C}^{-1}\mathbf{k})\big).$$

Proof (i) For given $\boldsymbol{\theta}$, we have

$$\begin{aligned}
\pi(\mathbf{y}|\boldsymbol{\theta}) &= \iint \pi(\mathbf{y}|\boldsymbol{\beta}, \sigma_f^2, \boldsymbol{\theta})\pi(\boldsymbol{\beta}, \sigma_f^2)d\boldsymbol{\beta}d\sigma_f^2 \\
&= \iint \pi(\mathbf{y}|\boldsymbol{\beta}, \sigma_f^2, \boldsymbol{\theta})\pi(\boldsymbol{\beta}|\sigma_f^2)\pi(\sigma_f^2)d\boldsymbol{\beta}d\sigma_f^2 \\
&= \frac{b^a}{\Gamma(a)(2\pi)^{n/2}\det(\mathbf{C})^{1/2}\det(\mathbf{T})^{1/2}\det(\mathbf{F}^T\mathbf{C}^{-1}\mathbf{F} + \mathbf{T}^{-1})^{1/2}} \\
&\quad \times \int \frac{\exp\left\{-\frac{b+\frac{RSS}{2}}{\sigma_f^2}\right\}}{(\sigma_f^2)^{n/2+a+1}}d\sigma_f^2 \\
&= \frac{\Gamma(\frac{n+2a}{2})b^a}{\Gamma(a)(2\pi)^{n/2}\det(\mathbf{C})^{1/2}\det(\mathbf{T})^{1/2}\det(\mathbf{F}^T\mathbf{C}^{-1}\mathbf{F} + \mathbf{T}^{-1})^{1/2}} \\
&\quad \times \left(b + \frac{RSS}{2}\right)^{-\frac{n+2a}{2}}
\end{aligned}$$

where

$$RSS = \mathbf{y}^T\left(\mathbf{C}^{-1} - \mathbf{C}^{-1}\mathbf{F}(\mathbf{F}^T\mathbf{C}^{-1}\mathbf{F} + \mathbf{T}^{-1})^{-1}\mathbf{F}^T\mathbf{C}^{-1}\right)\mathbf{y}.$$

As $\alpha \to \infty$, we have

$$\begin{aligned}
\pi(\boldsymbol{\theta}|\mathbf{y}) &= \lim_{\alpha\to\infty} \frac{\pi(y|\boldsymbol{\theta})\pi(\boldsymbol{\theta})}{\int \pi(\mathbf{y}|\boldsymbol{\theta})\pi(\boldsymbol{\theta})d\boldsymbol{\theta}} \\
&\propto \det(\mathbf{C})^{-1/2}\det(\mathbf{F}^T\mathbf{C}^{-1}\mathbf{F})^{-1/2}\left(b + \frac{1}{2}\mathbf{y}^T\boldsymbol{\Sigma}\mathbf{y}\right)^{-\frac{n+2a}{2}},
\end{aligned}$$

where $\Sigma = \mathbf{C}^{-1} - \mathbf{C}^{-1}\mathbf{F}(\mathbf{F}^T\mathbf{C}^{-1}\mathbf{F})^{-1}\mathbf{F}^T\mathbf{C}^{-1}$.

(ii) For given $\boldsymbol{\theta}$ and $\mathbf{y}$, we have

$$\begin{aligned}
\pi(z_*|\boldsymbol{\theta}, \mathbf{y}) &= \iint \pi(z_*|\mathbf{y}, \boldsymbol{\beta}, \sigma_f^2, \boldsymbol{\theta})\pi(\boldsymbol{\beta}, \sigma_f^2|\boldsymbol{\theta}, \mathbf{y})d\boldsymbol{\beta}d\sigma_f^2 \\
&= \iint \pi(z_*|\mathbf{y}, \boldsymbol{\beta}, \sigma_f^2, \boldsymbol{\theta})\pi(\boldsymbol{\beta}|\sigma_f^2, \boldsymbol{\theta}, \mathbf{y})\pi(\sigma_f^2|\boldsymbol{\theta}, \mathbf{y})d\boldsymbol{\beta}d\sigma_f^2,
\end{aligned}$$

where

$$z_*|\mathbf{y},\boldsymbol{\beta},\sigma_f^2,\boldsymbol{\theta} \sim \mathbb{N}\left(f(\mathbf{x}_*)^T\boldsymbol{\beta}+\mathbf{k}^T\mathbf{C}^{-1}(\mathbf{y}-\mathbf{F}\boldsymbol{\beta}), \sigma_f^2(1-\mathbf{k}^T\mathbf{C}^{-1}\mathbf{k})\right),$$
$$\boldsymbol{\beta}|\sigma_f^2,\boldsymbol{\theta},\mathbf{y} \sim \mathbb{N}(\hat{\boldsymbol{\beta}}, \sigma_f^2\boldsymbol{\Sigma}_{\hat{\boldsymbol{\beta}}}),$$
$$\sigma_f^2|\boldsymbol{\theta},\mathbf{y} \sim IG\left(a+\frac{n}{2}, b+\frac{RSS}{2}\right).$$

Then, it can be shown that

$$\pi(z_*|\boldsymbol{\theta},\mathbf{y}) = \frac{\Gamma\left(\frac{\nu+1}{2}\right)}{\Gamma\left(\frac{\nu}{2}\right)}\left(\frac{\lambda}{\pi\nu}\right)^{\frac{1}{2}}\left(1+\frac{\lambda(z_*-\mu)^2}{\nu}\right)^{-\frac{\nu+1}{2}},$$

when $\alpha \to \infty$.

The results in Proposition 5.2 are different from those obtained in [85] by using a noninformative prior on $\boldsymbol{\beta}$. For a special case where $\boldsymbol{\beta}$ and σ_f^2 are known a priori, we have the following corollary which will be exploited to derive a distributed implementation among sensor groups in Sect. 5.2.2.

Corollary 5.1 *In the case where $\boldsymbol{\beta}$ and σ_f^2 are known* a priori, (5.7) *and* (5.8) *can be simplified as*

$$\log w(\boldsymbol{\theta}|\mathbf{y}) = -\frac{1}{2}\log\det(\mathbf{C}) - \frac{1}{2}(\mathbf{y}-\mathbf{F}\boldsymbol{\beta})^T\mathbf{C}^{-1}(\mathbf{y}-\mathbf{F}\boldsymbol{\beta}),$$
$$z_*|\boldsymbol{\theta},\mathbf{y} \sim \mathcal{N}\left(f(\mathbf{x}_*)^T\boldsymbol{\beta}+\mathbf{k}^T\mathbf{C}^{-1}(\mathbf{y}-\mathbf{F}\boldsymbol{\beta}), \sigma_f^2(1-\mathbf{k}^T\mathbf{C}^{-1}\mathbf{k})\right).$$

If we draw m samples $\left\{\boldsymbol{\theta}^{(i)}\right\}_{i=1}^m$ from the prior distribution $\pi(\boldsymbol{\theta})$, the posterior predictive distribution in (5.6) can then be approximated by

$$\pi(z_*|\mathbf{y}) \approx \frac{\sum w(\boldsymbol{\theta}^{(i)}|\mathbf{y})\pi(z_*|\boldsymbol{\theta}^{(i)},\mathbf{y})}{\sum w(\boldsymbol{\theta}^{(i)}|\mathbf{y})}.$$

It follows that the predictive mean and variance can be obtained by

$$\mu_{z_*|\mathbf{y}} = \mathbb{E}(z_*|\mathbf{y}) \approx \frac{\sum w(\boldsymbol{\theta}^{(i)}|\mathbf{y})\mu_{z_*|\boldsymbol{\theta}^{(i)},\mathbf{y}}}{\sum w(\boldsymbol{\theta}^{(i)}|\mathbf{y})},$$
$$\sigma_{z_*|\mathbf{y}}^2 = \mathrm{Var}(z_*|\mathbf{y}) \approx \frac{\sum w(\boldsymbol{\theta}^{(i)}|\mathbf{y})\sigma_{z_*|\boldsymbol{\theta}^{(i)},\mathbf{y}}^2}{\sum w(\boldsymbol{\theta}^{(i)}|\mathbf{y})} + \frac{\sum w(\boldsymbol{\theta}^{(i)}|\mathbf{y})\left(\mu_{z_*|\boldsymbol{\theta}^{(i)},\mathbf{y}}-\mu_{z_*|\mathbf{y}}\right)^2}{\sum w(\boldsymbol{\theta}^{(i)}|\mathbf{y})},$$

where the mean and variance of the student's t-distribution $\pi(z_*|\boldsymbol{\theta}, \mathbf{y})$ are given by

$$\mu_{z_*|\boldsymbol{\theta},\mathbf{y}} = \mathbb{E}(z_*|\boldsymbol{\theta}, \mathbf{y}) = \mu,$$

$$\sigma^2_{z_*|\boldsymbol{\theta},\mathbf{y}} = \mathrm{Var}(z_*|\boldsymbol{\theta}, \mathbf{y}) = \frac{\tilde{a}}{\tilde{a}-1}\lambda.$$

5.1.4 Discrete Prior Distribution

To further reduce the computational demands from the Monte Carlo approach, we assign discrete uniform probability distributions to σ_s and σ_t as priors instead of continuous probability distributions. Assume that we know the range of parameters in $\boldsymbol{\theta}$, i.e.,

$$\sigma_s \in \left[\underline{\sigma}_s, \overline{\sigma}_s\right] \text{ and } \sigma_t \in \left[\underline{\sigma}_t, \overline{\sigma}_t\right],$$

where $\underline{\sigma}$ and $\overline{\sigma}$ denote the known lower-bound and upper-bound of the random variable σ, respectively. We constrain the possible choices of $\boldsymbol{\theta}$ on a finite set of grid points denoted by $\boldsymbol{\Theta}$. (Note here $\boldsymbol{\Theta}$ is defined as a finite set of discrete points, not to be confused with the continuous space definition used in previous chapters.) Hence, $\pi(\boldsymbol{\theta})$ is now a probability mass function (i.e., $\sum_{\boldsymbol{\theta}\in\boldsymbol{\Theta}} \pi(\boldsymbol{\theta}) = 1$) as opposed to a probability density. The integration in (5.6) is reduced to the following summation

$$\pi(z_*|\mathbf{y}) = \sum_{\boldsymbol{\theta}\in\boldsymbol{\Theta}} \pi(z_*|\boldsymbol{\theta}, \mathbf{y})\pi(\boldsymbol{\theta}|\mathbf{y}), \tag{5.9}$$

where the posterior distribution of $\boldsymbol{\theta}$ is evaluated on the grid points in $\boldsymbol{\theta}$ by

$$\pi(\boldsymbol{\theta}|\mathbf{y}) = \frac{w(\boldsymbol{\theta}|\mathbf{y})\pi(\boldsymbol{\theta})}{\sum_{\boldsymbol{\theta}\in\boldsymbol{\theta}} w(\boldsymbol{\theta}|\mathbf{y})\pi(\boldsymbol{\theta})}. \tag{5.10}$$

In order to obtain the posterior predictive distribution in (5.9), the computation of $\pi(z_*|\boldsymbol{\theta}, \mathbf{y})$ and $w(\boldsymbol{\theta}|\mathbf{y})$ for all $\boldsymbol{\theta} \in \boldsymbol{\Theta}$ using the results from Proposition 5.2 (or Corollary 5.1 for a special case) are necessary. Note that these quantities are available in closed-form which reduces the computational burden significantly.

5.2 Sequential Bayesian Prediction

Although the aforementioned efforts in Sects. 5.1.3 and 5.1.4 reduce the computational cost significantly, the number of observations (that mobile sensing agents collect) n increases with the time t. For each $\boldsymbol{\theta} \in \boldsymbol{\Theta}$, an $n \times n$ positive definite matrix $\mathbf{C}$ needs to be inverted which requires time $O(n^3)$ using standard methods.

This motivates us to design scalable sequential Bayesian prediction algorithms by using subsets of observations.

5.2.1 Scalable Bayesian Prediction Algorithm

The computation of $\pi(z_*|\mathbf{y}_{1:t})$ soon becomes infeasible as t increases. To overcome this drawback while maintaining the Bayesian framework, we propose to use subsets of all observations $\mathbf{y}_{1:t} \in \mathbb{R}^n$. However, instead of using truncated local observations only as in [2], Bayesian inference will be drawn based on two sets of observations:

- First, a set of local observations near target points $\tilde{\mathbf{y}}$ which will improve the quality of the prediction, and
- second, a cumulative set of observations $\bar{\mathbf{y}}$ which will minimize the uncertainty in the estimated parameters.

Taken together, they improve the quality of prediction as the number of observations increases. We formulate this idea in detail in the following paragraph. For notational simplicity, we define $\mathbf{y} \in \mathbb{R}^n$ as a subset of all observations $\mathbf{y}_{1:t}$ which will be used for Bayesian prediction. We partition $\mathbf{y}$ into two subsets, namely $\bar{\mathbf{y}}$ and $\tilde{\mathbf{y}}$. Let $\bar{\mathbf{F}}$ and $\tilde{\mathbf{F}}$ be the counterparts of $\mathbf{F}$ defined in (5.2) for $\bar{\mathbf{y}}$ and $\tilde{\mathbf{y}}$, respectively. The following lemma provides the conditions under which any required function of $\mathbf{y}$ in Proposition 5.2 can be decoupled.

Lemma 5.1 *For a given* $\theta \in \boldsymbol{\theta}$, *let* $\mathbf{C} = \mathrm{Corr}(\mathbf{y}, \mathbf{y}|\theta)$, $\bar{\mathbf{C}} = \mathrm{Corr}(\bar{\mathbf{y}}, \bar{\mathbf{y}}|\theta)$, $\tilde{\mathbf{C}} = \mathrm{Corr}(\tilde{\mathbf{y}}, \tilde{\mathbf{y}}|\theta)$, $\mathbf{k} = \mathrm{Corr}(\mathbf{y}, z_*|\theta)$, $\bar{\mathbf{k}} = \mathrm{Corr}(\bar{\mathbf{y}}, z_*|\theta)$, *and* $\tilde{\mathbf{k}} = \mathrm{Corr}(\tilde{\mathbf{y}}, z_*|\theta)$. *If the following conditions are satisfied*

C1: $\mathrm{Corr}(\tilde{\mathbf{y}}, \bar{\mathbf{y}}|\theta) = 0$, *i.e.,* $\tilde{\mathbf{y}}$ *and* $\bar{\mathbf{y}}$ *are uncorrelated, and*
C2: $\mathrm{Corr}(\bar{\mathbf{y}}, z_*|\theta) = 0$, *i.e.,* $\bar{\mathbf{y}}$ *and* z_* *are uncorrelated,*

then we have the following results:

$$
\begin{aligned}
&\mathbf{F}^T\mathbf{C}^{-1}\mathbf{F} = \bar{\mathbf{F}}^T\bar{\mathbf{C}}^{-1}\bar{\mathbf{F}} + \tilde{\mathbf{F}}^T\tilde{\mathbf{C}}^{-1}\tilde{\mathbf{F}} \in \mathbb{R}^{p\times p},\\
&\mathbf{F}^T\mathbf{C}^{-1}\mathbf{y} = \bar{\mathbf{F}}^T\bar{\mathbf{C}}^{-1}\bar{\mathbf{y}} + \tilde{\mathbf{F}}^T\tilde{\mathbf{C}}^{-1}\tilde{\mathbf{y}} \in \mathbb{R}^{p},\\
&\mathbf{y}^T\mathbf{C}^{-1}\mathbf{y} = \bar{\mathbf{y}}^T\bar{\mathbf{C}}^{-1}\bar{\mathbf{y}} + \tilde{\mathbf{y}}^T\tilde{\mathbf{C}}^{-1}\tilde{\mathbf{y}} \in \mathbb{R},\\
&\log\det\mathbf{C} = \log\det\bar{\mathbf{C}} + \log\det\tilde{\mathbf{C}} \in \mathbb{R},\\
&\mathbf{F}^T\mathbf{C}^{-1}\mathbf{k} = \tilde{\mathbf{F}}^T\tilde{\mathbf{C}}^{-1}\tilde{\mathbf{k}} \in \mathbb{R}^{p},\\
&\mathbf{k}^T\mathbf{C}^{-1}\mathbf{k} = \tilde{\mathbf{k}}^T\tilde{\mathbf{C}}^{-1}\tilde{\mathbf{k}} \in \mathbb{R}.
\end{aligned}
$$

Proof The results follow by noting the correlation matrix $\mathbf{C}$ can be decoupled such that $\mathbf{C} = \mathrm{diag}(\bar{\mathbf{C}}, \tilde{\mathbf{C}})$ and $\bar{\mathbf{k}} = \mathbf{0}$.

Remark 5.2 In order to compute the posterior predictive distribution $\pi(z_*|\mathbf{y})$ (or the predictive mean and variance) in (5.9), $\pi(z_*|\boldsymbol{\theta}, \mathbf{y})$ and $\pi(\boldsymbol{\theta}|\mathbf{y})$ for all $\boldsymbol{\theta} \in \boldsymbol{\Theta}$ need to be calculated. Notice that the posterior distribution of $\boldsymbol{\theta}$ can be obtained by computing $w(\boldsymbol{\theta}|\mathbf{y})$ in (5.7). Suppose $\bar{\mathbf{F}}^T\bar{\mathbf{C}}^{-1}\bar{\mathbf{F}} \in \mathbb{R}^{p\times p}$, $\bar{\mathbf{F}}^T\bar{\mathbf{C}}^{-1}\bar{\mathbf{y}} \in \mathbb{R}^p$, $\bar{y}^T\bar{\mathbf{C}}^{-1}\bar{\mathbf{y}} \in \mathbb{R}$, and $\log\det\bar{\mathbf{C}} \in \mathbb{R}$ are known for all $\boldsymbol{\theta} \in \boldsymbol{\Theta}$. If $\tilde{\mathbf{F}}^T\tilde{\mathbf{C}}^{-1}\tilde{\mathbf{F}} \in \mathbb{R}^{p\times p}$, $\tilde{\mathbf{F}}^T\tilde{\mathbf{C}}^{-1}\tilde{\mathbf{y}} \in \mathbb{R}^p$, $\tilde{\mathbf{y}}^T\tilde{\mathbf{C}}^{-1}\tilde{\mathbf{y}} \in \mathbb{R}$, and $\log\det\tilde{\mathbf{C}} \in \mathbb{R}$ for all $\boldsymbol{\theta} \in \boldsymbol{\Theta}$ have fixed computation times, then (5.7) and (5.8) can be computed in constant time due to decoupling results of Lemma 5.1.

The following theorem provides a way to design scalable sequential Bayesian prediction algorithms.

Theorem 5.1 *Consider the discrete prior probability $\pi(\boldsymbol{\theta})$ and the compactly supported kernel function $\phi_t(\cdot)$. If we select $\eta \geq \lfloor\bar{\sigma}_t\rfloor \in \mathbb{Z}_{>0}$, $\Delta \in \mathbb{Z}_{>0}$ and define*

$$
\begin{aligned}
c_t &:= \max\left(\left\lfloor\frac{t-\Delta}{\Delta+\eta}\right\rfloor, 0\right) \in \mathbb{R},\\
\boldsymbol{\xi}_j &:= \mathbf{y}_{(j-1)(\Delta+\eta)+1:(j-1)(\Delta+\eta)+\Delta} \in \mathbb{R}^{\Delta N},\\
\bar{\mathbf{y}} &:= (\boldsymbol{\xi}_1^T, \ldots, \boldsymbol{\xi}_{c_t}^T)^T \in \mathbb{R}^{\Delta N c_t},\\
\tilde{\mathbf{y}} &:= \mathbf{y}_{t-\Delta+1:t} \in \mathbb{R}^{\Delta N},
\end{aligned} \tag{5.11}
$$

where $\lfloor\cdot\rfloor$ is the floor function defined by $\lfloor x\rfloor := \max\{m \in \mathbb{Z} \,|\, m \leq x\}$, then the posterior predictive distribution in (5.9) *can be computed in constant time (i.e., does not grow with the time t).*

Proof By construction, conditions $C1$–2 in Lemma 5.1 are satisfied. Hence, it follows from Remark 5.2 that the posterior predictive distribution can be computed in constant time.

Remark 5.3 In Theorem 5.1, $\eta \geq \lfloor\bar{\sigma}_t\rfloor$ guarantees the time distance between $\boldsymbol{\xi}_j$ and $\boldsymbol{\xi}_{j+1}$ is large enough such that the conditions in Lemma 5.1 are satisfied. Notice that Δ is a tuning parameter for users to control the trade-off between the prediction quality and the computation efficiency. A large value for Δ yields a small predictive variance but long computation time, and vice versa. An illustrative example with three agents sampling the spatiotemporal Gaussian process in 1-D space is shown in Fig. 5.1.

Based on Theorem 5.1, we provide the centralized sequential Bayesian prediction algorithm as shown in Table 5.2.

5.2.2 Distributed Implementation for a Special Case

In this subsection, we will show a distributed way (among agent groups) to implement the proposed algorithm for a special case in which $\boldsymbol{\beta}$ and σ_f^2 are assumed to be known

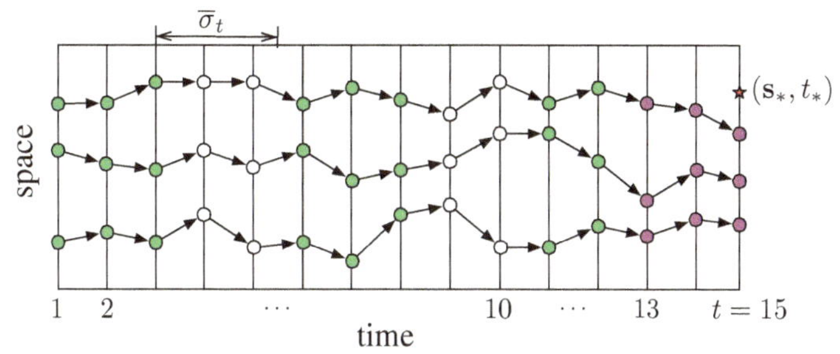

Fig. 5.1 Example with three agents sampling the spatiotemporal Gaussian process in 1-D space and performing Bayesian inference. In this example, $\overline{\sigma}_t = 2.5$, $\eta = 2$, $\Delta = 3$, $t = 15$, $c_t = 2$, $\bar{\mathbf{y}} = (\mathbf{y}_{1:3}^T, \mathbf{y}_{6:8}^T)^T$ and $\tilde{\mathbf{y}} = \mathbf{y}_{13:15}$

a priori. The assumption for this special case is the exact opposite of the one made in [46] where β and σ_f^2 are unknown and $\boldsymbol{\theta}$ is known *a priori*.

To develop a distributed scheme among agent groups for data fusion in Bayesian statistics, we exploit the compactly supported kernel for space. Let $C_s(h)$ in (5.1) also be a compactly supported kernel function as $C_t(h)$ so that the correlation vanishes when the spatial distance between two inputs is larger than σ_s, i.e., $C_s(h) = 0, \forall h > 1$.

Consider a case in which M groups of spatially distributed agents sample a spatiotemporal Gaussian process over a large region $\mathcal{Q}$. Each group is in charge of its sub-region of $\mathcal{Q}$. The identity of each group is indexed by $\mathcal{V} := \{1, \ldots, M\}$. Each agent in group i is indexed by $\mathcal{I}^{[i]} := \{1, \ldots, N\}$. The leader of group i is referred to as leader i, which implements the centralized scheme to make prediction on its sub-region using local observations and the globally updated posterior distribution of $\boldsymbol{\theta}$. Therefore, the posterior distribution of $\boldsymbol{\theta}$ shall be updated correctly using all observations from all groups (or agents) in a distributed fashion.

Let $\mathcal{G}(t) := (\mathcal{V}, \mathcal{E}(t))$ be an undirected communication graph such that an edge $(i, j) \in \mathcal{E}(t)$ if and only if leader i can communicate with leader j at time t. We define the neighborhood of leader i at time t by $\mathcal{N}_i(t) := \{j \in \mathcal{V} \mid (i, j) \in \mathcal{E}(t), j \neq i\}$. Let $a^{[i]}$ denote the quantity as a in the centralized scheme for group i. We then have the following theorem.

Theorem 5.2 *Assume that $\bar{\mathbf{y}}^{[i]}$ and $\tilde{\mathbf{y}}^{[i]}$ for leader i are selected accordingly to Theorem 5.1 in time-wise. Let $\tilde{\mathbf{y}}$ defined by $\tilde{\mathbf{y}} := ((\tilde{\mathbf{y}}^{[1]})^T, \ldots, (\tilde{\mathbf{y}}^{[M]})^T)^T$. If the following condition is satisfied*

C3: $\|\mathbf{q}_\ell^{[i]}(t) - \mathbf{q}_\nu^{[j]}(t')\| \geq \overline{\sigma}_s, \forall i \neq j, \forall \ell \in \mathcal{I}^{[i]}, \forall \nu \in \mathcal{I}^{[j]}$,

in the spatial domain, then the weights $w(\boldsymbol{\theta}|\mathbf{y})$, based on all observations from all agents, can be obtained from

$$\log w(\boldsymbol{\theta}|\mathbf{y}) = \log w(\boldsymbol{\theta}|\bar{\mathbf{y}}) + \sum_{i=1}^{M} \log w(\boldsymbol{\theta}|\tilde{\mathbf{y}}^{[i]}). \tag{5.12}$$

Table 5.2 Centralized Bayesian prediction algorithm

Input:

(1) prior distribution on σ_f^2, *i.e.*, $\pi(\sigma_f^2) = IG(a, b)$
(2) prior distribution on $\boldsymbol{\theta} \in \boldsymbol{\Theta}$, *i.e.*, $\pi(\boldsymbol{\theta})$
(3) tuning variables Δ and η
(4) number of agents N
(5) $\mathcal{M}(\boldsymbol{\theta}).\mathbf{A} = \mathbf{0} \in \mathbb{R}^{p \times p}$, $\mathcal{M}(\boldsymbol{\theta}).B = 0 \in \mathbb{R}$, $\mathcal{M}(\boldsymbol{\theta}).\mathbf{c} = \mathbf{0} \in \mathbb{R}^p$, $\mathcal{M}(\boldsymbol{\theta}).D = 0 \in \mathbb{R}$, $\mathcal{M}_0(\boldsymbol{\theta}) = \mathcal{M}(\boldsymbol{\theta})$, $\forall \boldsymbol{\theta} \in \boldsymbol{\Theta}$

Output:

(1) The predictive mean at location $\mathbf{s}_* \in \mathcal{S}$ and time $t_* = t$, *i.e.*, $\mu_{z_*|\mathbf{y}}$
(2) The predictive variance at location $\mathbf{s}_* \in \mathcal{S}$ and time $t_* = t$, *i.e.*, $\sigma^2_{z_*|\mathbf{y}}$

At time t, the central station does:

1: receive observations $\mathbf{y}_t$ from agents, set $\tilde{\mathbf{y}} = \mathbf{y}_{t-\Delta+1:t}$ and $n = N\Delta$
2: compute $\tilde{\mathbf{F}} = (f(\tilde{\mathbf{x}}^{(1)}), \cdots, f(\tilde{\mathbf{x}}^{(n)}))^T$ where $\tilde{\mathbf{x}}^{(i)}$ is the input of the i-th element in $\tilde{\mathbf{y}}$
3: **for** each $\boldsymbol{\theta} \in \boldsymbol{\theta}$ **do**
4: compute $\tilde{\mathbf{C}} = \mathrm{Cor}(\tilde{\mathbf{y}}, \tilde{\mathbf{y}}) \in \mathbb{R}^{n \times n}$
5: compute the key values

$$\mathbf{F}^T\mathbf{C}^{-1}\mathbf{F} = \mathcal{M}(\boldsymbol{\theta}).\mathbf{A} + \tilde{\mathbf{F}}^T\tilde{\mathbf{C}}^{-1}\tilde{\mathbf{F}} \in \mathbb{R}^{p \times p},\ \mathbf{y}^T\mathbf{C}^{-1}\mathbf{y} = \mathcal{M}(\boldsymbol{\theta}).B + \tilde{\mathbf{y}}^T\tilde{\mathbf{C}}^{-1}\tilde{\mathbf{y}} \in \mathbb{R},$$
$$\mathbf{F}^T\mathbf{C}^{-1}\mathbf{y} = \mathcal{M}(\boldsymbol{\theta}).\mathbf{c} + \tilde{\mathbf{F}}^T\tilde{\mathbf{C}}^{-1}\tilde{\mathbf{y}} \in \mathbb{R}^p,\ \log\det\mathbf{C} = \mathcal{M}(\boldsymbol{\theta}).D + \log\det\tilde{\mathbf{C}} \in \mathbb{R}$$

6: compute $\tilde{a} = a + \frac{n}{2}$ and

$$\tilde{b} = b + \tfrac{1}{2}\mathbf{y}^T\mathbf{C}^{-1}\mathbf{y} - \tfrac{1}{2}(\mathbf{F}^T\mathbf{C}^{-1}\mathbf{y})^T(\mathbf{F}^T\mathbf{C}^{-1}\mathbf{F})^{-1}(\mathbf{F}^T\mathbf{C}^{-1}\mathbf{y})$$

7: update weights via

$$\log w(\boldsymbol{\theta}|\mathbf{y}) = -\tfrac{1}{2}\log\det\mathbf{C} - \tfrac{1}{2}\log\det(\mathbf{F}^T\mathbf{C}^{-1}\mathbf{F}) - \tilde{a}\log\tilde{b}$$

8: **for** each $\mathbf{s}_* \in \mathcal{S}$ **do**
9: compute $f(\mathbf{x}_*) \in \mathbb{R}^p$, $\tilde{\mathbf{k}} = \mathrm{Corr}(\tilde{\mathbf{y}}, z_*) \in \mathbb{R}^n$
10: compute predictive mean and variance for given $\boldsymbol{\theta}$

$$\mu_{z_*|\boldsymbol{\theta},\mathbf{y}} = \tilde{\mathbf{k}}\tilde{\mathbf{C}}^{-1}\tilde{\mathbf{y}} + (f(\mathbf{x}_*) - \tilde{\mathbf{F}}^T\tilde{\mathbf{C}}^{-1}\tilde{\mathbf{k}})^T(\mathbf{F}^T\mathbf{C}^{-1}\mathbf{F})^{-1}(\mathbf{F}^T\mathbf{C}^{-1}\mathbf{y}),$$
$$\sigma^2_{z_*|\boldsymbol{\theta},\mathbf{y}} = \frac{\tilde{b}}{\tilde{a}-1}\left((1 - \tilde{\mathbf{k}}^T\tilde{\mathbf{C}}^{-1}\tilde{\mathbf{k}}) + (f(\mathbf{x}_*) - \tilde{\mathbf{F}}^T\tilde{\mathbf{C}}^{-1}\tilde{\mathbf{k}})^T(\mathbf{F}^T\mathbf{C}^{-1}\mathbf{F})^{-1}(f(\mathbf{x}_*) - \tilde{\mathbf{F}}^T\tilde{\mathbf{C}}^{-1}\tilde{\mathbf{k}})\right)$$

11: **end for**
12: **if** $\mathrm{mod}(t, \Delta + \eta) = \Delta$ **then**
13: set $\mathcal{M}(\boldsymbol{\theta}) = \mathcal{M}_0(\boldsymbol{\theta})$, then $\mathcal{M}_0(\boldsymbol{\theta}).\mathbf{A} = \mathbf{F}^T\mathbf{C}^{-1}\mathbf{F}$, $\mathcal{M}_0(\boldsymbol{\theta}).B = \mathbf{y}^T\mathbf{C}^{-1}\mathbf{y}$, $\mathcal{M}_0(\boldsymbol{\theta}).\mathbf{c} = \mathbf{F}^T\mathbf{C}^{-1}\mathbf{y}$, and $\mathcal{M}_0(\boldsymbol{\theta}).D = \log\det\mathbf{C}$
14: **end if**
15: **end for**
16: compute the posterior distribution

$$\pi(\boldsymbol{\theta}|\mathbf{y}) = \frac{w(\boldsymbol{\theta}|\mathbf{y})\pi(\boldsymbol{\theta})}{\sum_{\boldsymbol{\theta}} w(\boldsymbol{\theta}|\mathbf{y})\pi(\boldsymbol{\theta})}$$

17: compute the predictive mean and variance

$$\mu_{z_*|\mathbf{y}} = \sum_{\boldsymbol{\theta}} \mu_{z_*|\boldsymbol{\theta},\mathbf{y}}\pi(\boldsymbol{\theta}|\mathbf{y}),$$
$$\sigma^2_{z_*|\mathbf{y}} = \sum_{\boldsymbol{\theta}} \sigma^2_{z_*|\boldsymbol{\theta},\mathbf{y}}\pi(\boldsymbol{\theta}|\mathbf{y}) + \sum_{\boldsymbol{\theta}} \left(\mu_{z_*|\boldsymbol{\theta},\mathbf{y}} - \mu_{z_*|\mathbf{y}}\right)^2 \pi(\boldsymbol{\theta}|\mathbf{y}).$$

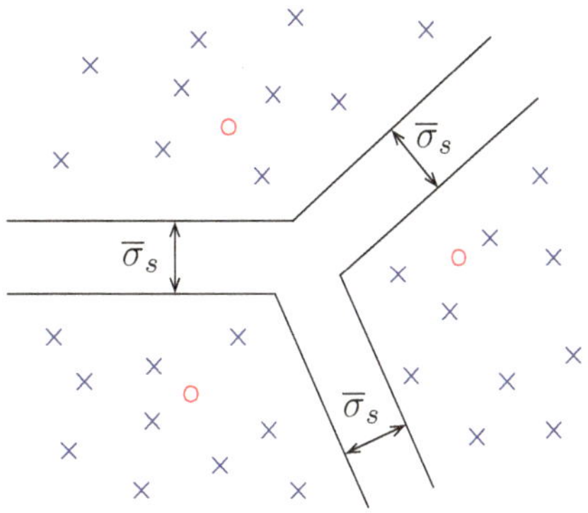

Fig. 5.2 Example with three group of agents sampling the spatiotemporal Gaussian process in 2-D space and performing Bayesian prediction. The symbol 'o' denotes the position of a leader for a group and the symbol 'x' denotes the position of an agent. Distance between any two sub-regions is enforced to be greater than $\overline{\sigma}_s$ which enables the distributed Bayesian prediction

Proof The result follows by noting $\mathrm{Corr}(\tilde{\mathbf{y}}^{[i]}, \tilde{\mathbf{y}}^{[j]}|\boldsymbol{\theta}) = 0, \forall i \neq j$, when the condition $C3$ is satisfied.

An exemplary configuration of agents which satisfies $C3$ is shown in Fig. 5.2.

Suppose that the communication graph $\mathcal{G}(t)$ is connected for all time t. Then the average $\frac{1}{M}\sum_{i=1}^{M} \log w(\boldsymbol{\theta}|\tilde{\mathbf{y}}^{[i]})$ can be achieved asymptotically via discrete-time average-consensus algorithm [110]:

$$\log w(\boldsymbol{\theta}|\tilde{\mathbf{y}}^{[i]}) \leftarrow \log w(\boldsymbol{\theta}|\tilde{\mathbf{y}}^{[i]}) + \epsilon \sum_{j \in \mathcal{N}_i} \left(\log w(\boldsymbol{\theta}|\tilde{\mathbf{y}}^{[j]}) - \log w(\boldsymbol{\theta}|\tilde{\mathbf{y}}^{[i]}) \right),$$

with $0 < \epsilon < 1/\Delta(G)$ that depends on the maximum node degree of the network $\Delta(G) = \max_i |\mathcal{N}_i|$.

5.2.3 Adaptive Sampling

At time t, the goal of the navigation of agents is to improve the quality of prediction of the field $\mathcal{Q}$ at the next sampling time $t+1$. Therefore, mobile agents should move to the most informative sampling locations $\mathbf{q}(t+1) = (\mathbf{q}_1(t+1)^T, \ldots, \mathbf{q}_N(t+1)^T)^T$ at time $t+1$ in order to reduce the prediction error [44].

Suppose at time $t+1$, agents move to a new set of positions $\tilde{\mathbf{q}} = (\tilde{\mathbf{q}}_1^T, \ldots, \tilde{\mathbf{q}}_N^T)^T$. The mean squared prediction error is defined as

$$J(\tilde{\mathbf{q}}) = \int_{\mathbf{s} \in \mathcal{S}} \mathbb{E}\left[(z(\mathbf{s}, t+1) - \hat{z}(\mathbf{s}, t+1))^2 \right] d\mathbf{s}, \tag{5.13}$$

where $\hat{z}(\mathbf{s}, t+1)$ is obtained as in (5.9). Due to the fact that $\boldsymbol{\theta}$ has a distribution, the evaluation of (5.13) becomes computationally prohibitive. To simplify the optimization, we propose to utilize a maximum *a posteriori* (MAP) estimate of $\boldsymbol{\theta}$ at time t, denoted by $\hat{\boldsymbol{\theta}}_t$, i.e.,

$$\hat{\boldsymbol{\theta}}_t = \arg\max_{\boldsymbol{\theta} \in \boldsymbol{\theta}} \pi(\boldsymbol{\theta}|\mathbf{y}),$$

where $\mathbf{y}$ is the subset of all observations used up to time t. The next sampling positions can be obtained by solving the following optimization problem

$$\mathbf{q}(t+1) = \arg\min_{\tilde{\mathbf{q}}_i \subset \mathcal{Q}} \int_{\mathbf{s} \in \mathcal{S}} \mathrm{Var}(z(\mathbf{s}, t+1)|\mathbf{y}, \hat{\boldsymbol{\theta}}_t) d\mathbf{s}. \tag{5.14}$$

This problem can be solved using standard constrained nonlinear optimization techniques (e.g., the conjugate gradient algorithm), possibly taking into account mobility constraints of mobile sensors.

Remark 5.4 The proposed control algorithm in (5.14) is truly adaptive in the sense that the new sampling positions are functions of all collected observations. On the other hand, if all parameters are known, the optimization in (5.14) can be performed offline without taking any measurements.

5.3 Simulation

In this section, we apply the proposed sequential Bayesian prediction algorithms to spatiotemporal Gaussian processes with a correlation function in Sect. 5.1. The Gaussian process was numerically generated through circulant embedding of the covariance matrix for the simulation study [107]. This technique allows us to numerically generate a large number of realizations of the Gaussian process.

5.3.1 MCMC-Based Approach on a 1-D Scenario

We consider a scenario in which $N = 5$ agents sample the spatiotemporal Gaussian process in 1-D space and the central station performs Bayesian prediction. The surveillance region $\mathcal{Q}$ is given by $\mathcal{Q} = [0, 10]$. We consider the squared exponential function

$$C_s(h) = \exp\left(-\frac{1}{2}h^2\right),$$

for space correlation and a compactly supported correlation function [111] for time as

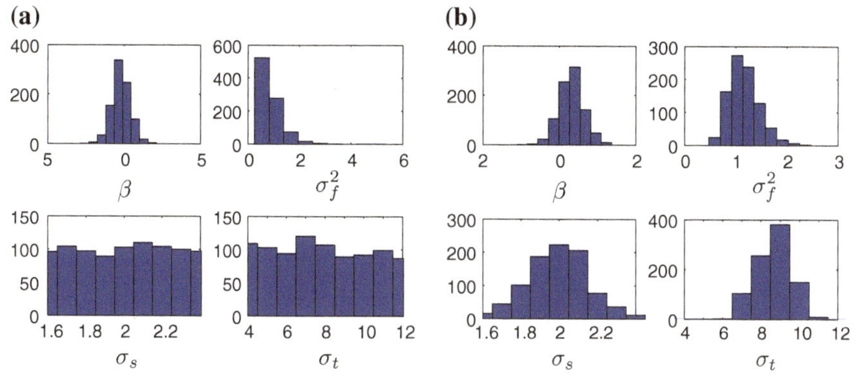

Fig. 5.3 Posterior distribution of β, σ_f^2, σ_s, and σ_t at **a** $t = 1$, and **b** $t = 20$

$$C_t(h) = \begin{cases} \frac{(1-h)\sin(2\pi h)}{2\pi h} + \frac{1-\cos(2\pi h)}{\pi \times 2\pi h}, & 0 \leq h \leq 1, \\ 0, & \text{otherwise,} \end{cases} \tag{5.15}$$

The signal-to-noise ratio γ is set to be 26 dB which corresponds to $\sigma_w = 0.158$. The true values for the parameters used in simulating the Gaussian process are given by $(\beta, \sigma_f^2, \sigma_s, \sigma_t) = (0, 1, 2, 8)$. Notice that the mean function is assumed to be an unknown random variable, i.e., the dimension of the regression coefficient β is 1. We assume that $\beta|\sigma_f^2$ has the noninformative prior and $\sigma_f^2 \sim IG(3, 20)$. The Gibbs sampler in Table 5.1 was used to generate samples from the posterior distribution of the parameters. A random sampling strategy was used in which agents make observations at random locations at each time $t \in \mathbb{Z}_{>0}$. The prediction was evaluated at each time step for 51 uniform grid points within $\mathcal{Q}$.

The histograms of the samples at time $t = 1$ and $t = 10$ are shown in Fig. 5.3a and Fig. 5.3b, respectively. It is clear that the distributions of the parameters are centered around the true values with 100 observations at time $t = 20$. The prediction results at time $t = 1$ and $t = 20$ are shown in Fig. 5.4a and Fig. 5.4b, respectively. However, with only 100 observations, the running time using the full Bayesian approach is about several minutes which will soon become intractable.

5.3.2 Centralized Scheme on 1-D Scenario

We consider the same scenario in which $N = 5$ agents sample the spatiotemporal Gaussian process in 1-D space and the central station performs Bayesian prediction. The true values for the parameters used in simulating the Gaussian process are given by $(\beta, \sigma_f^2, \sigma_s, \sigma_t) = (20, 10, 2, 8)$. Notice that the mean function is assumed to be an unknown random variable, i.e., the dimension of the regression coefficient β is 1. We assume that $\beta|\sigma_f^2$ has the noninformative prior and $\sigma_f^2 \sim IG(3, 20)$.

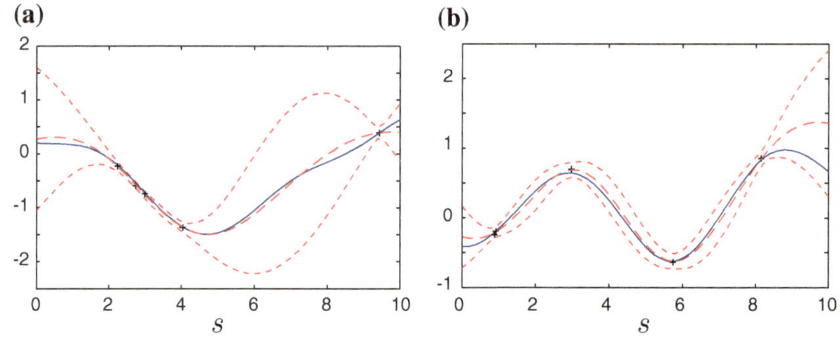

Fig. 5.4 Prediction at **a** $t = 1$, and **b** $t = 20$ using the MCMC-based approach. The true fields are plotted in *blue solid lines*. The predicted fields are plotted in *red dash-dotted lines*. The area between *red dotted lines* indicates the 95 % confidence interval

We also assume the bounds of $\boldsymbol{\theta}$, viz. $\sigma_s \in [1.6, 2.4]$ and $\sigma_t \in [4, 12]$ are known. $\Delta = 12$ is used and $\eta = 11$ is selected satisfying the condition in Theorem 5.1. We use a discrete uniform probability distribution for $\pi(\boldsymbol{\theta})$ as shown in Fig. 5.6a. The adaptive sampling strategy was used in which agents make observations at each time $t \in \mathbb{Z}_{>0}$. The prediction was evaluated at each time step for 51 uniform grid points within $\mathcal{Q}$.

Figure 5.5 shows the comparison between predictions at time $t = 1$ using (a) the maximum likelihood (ML) based approach, and (b) the proposed fully Bayesian approach. The ML based approach first generates a point estimate of the hyperparameters and then uses them as true ones for computing the prediction and the prediction error variance. In this simulation, a poor point estimate on $\boldsymbol{\theta}$ was achieved by maximizing the likelihood function. As a result, the prediction and the associated prediction error variance are incorrect and are far from being accurate for a small number of observations. On the other hand, the fully Bayesian approach which incorporates the prior knowledge of $\boldsymbol{\theta}$ and uncertainties in $\boldsymbol{\theta}$ provides a more accurate prediction and an exact confidence interval.

Using the proposed sequential Bayesian prediction algorithm along with the adaptive sampling strategy, the prior distribution was updated in a sequential manner. At time $t = 100$, the posterior distribution of $\boldsymbol{\theta}$ is shown in Fig. 5.6b. With a larger number of observations, the support for the posterior distribution of $\boldsymbol{\theta}$ becomes smaller and the peak gets closer to the true value. As shown in Fig. 5.7a, the quality of the prediction at time $t = 100$ is significantly improved. At time $t = 300$, the prior distribution was further updated which is shown in Fig. 5.6c. At this time, $\boldsymbol{\theta} = (2, 8)^T$, which is the true value, has the highest probability. The prediction is also shown in Fig. 5.7b. This demonstrates the usefulness and correctness of our algorithm. The running time at each time step is fixed, which is around 12s using Matlab, R2008a (MathWorks) in a PC (2.4GHz Dual-Core Processor).

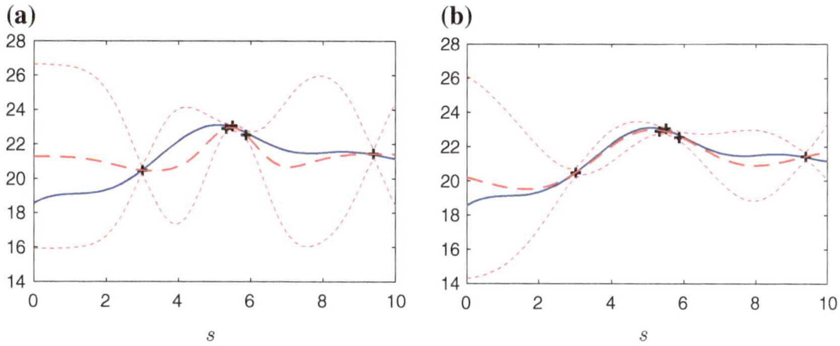

Fig. 5.5 Prediction at $t = 1$ using (**a**) the maximum likelihood based approach, and (**b**) the proposed fully Bayesian approach. The true fields are plotted in *blue solid lines*. The predicted fields are plotted in *red dash-dotted lines*. The area between *red dotted lines* indicates the 95 % confidence interval

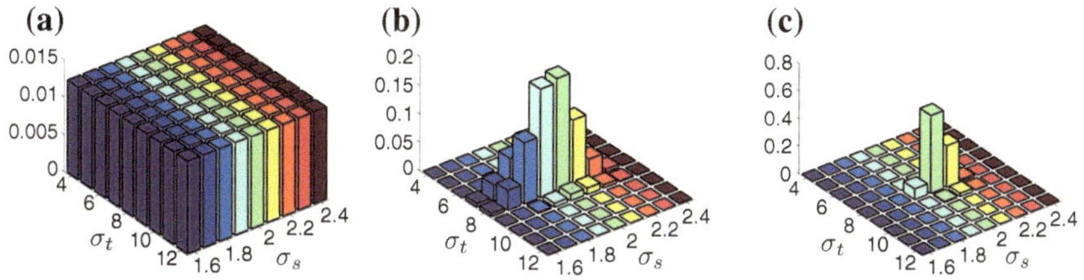

Fig. 5.6 **a** Prior distribution $\boldsymbol{\theta}$, **b** posterior distribution of $\boldsymbol{\theta}$ at time $t = 100$, **c** posterior distribution of $\boldsymbol{\theta}$ at time $t = 300$

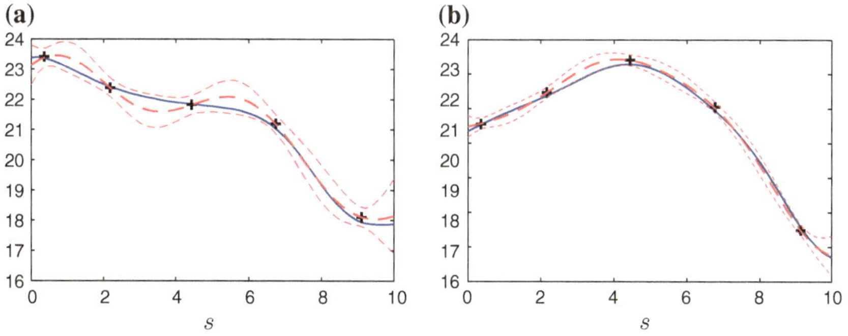

Fig. 5.7 Prediction at **a** $t = 100$, and **b** $t = 300$ using the centralized sequential Bayesian approach. The true fields are plotted in *blue solid lines*. The predicted fields are plotted in *red dash-dotted lines*. The area between *red dotted lines* indicates the 95 % confidence interval

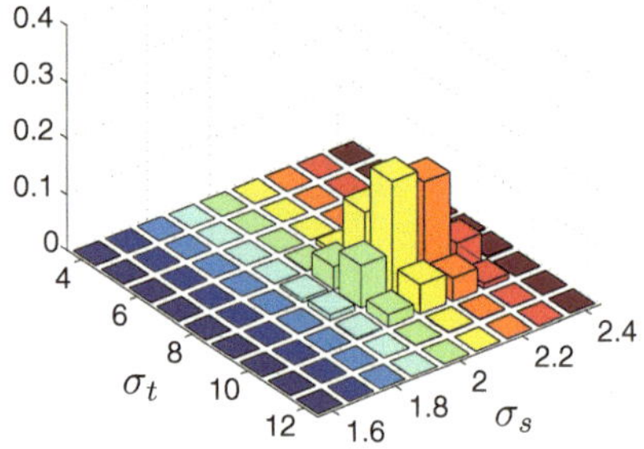

Fig. 5.8 Posterior distribution of $\boldsymbol{\theta}$ at time $t = 100$ using the distributed algorithm

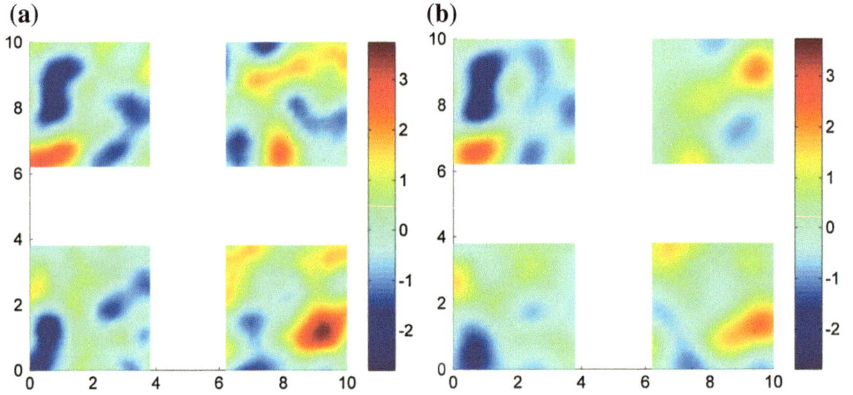

Fig. 5.9 Comparison of (**a**) the true field at $t = 100$ and (**b**) the predicted field at $t = 100$ using the distributed algorithm

5.3.3 Distributed Scheme on 2-D Scenario

Finally, we consider a scenario in which there are 4 groups, each of which contain 10 agents sampling the spatiotemporal Gaussian process in 2-D space. The surveillance region $\mathcal{Q}$ is given by $\mathcal{Q} = [0, 10] \times [0, 10]$. The parameter values used in simulating the Gaussian process are given by $\boldsymbol{\theta} = (\sigma_s, \sigma_t)^T = (2, 8)^T$, $\beta = 0$, and $\sigma_f^2 = 1$, last two values of which are assumed to be known a priori. To use the distributed scheme, we only consider compactly supported kernel functions for both space and time. In particular, we consider $C_s(h) = C_t(h)$ as in (5.15). We also assume the fact that $\sigma_s \in [1.6, 2.4]$ and $\sigma_t \in [4, 12]$ are known a priori. $\Delta = 12$ is used and $\eta = 11$ is selected satisfying the condition in Theorem 5.1. The region $\mathcal{Q}$ is divided into 4 square sub-regions with equal size areas as shown in Fig. 5.9a. Distance between any two sub-regions is enforced to be greater than $\bar{\sigma}_s = 2.4$, satisfying the condition in

Theorem 5.2, which enables the distributed Bayesian prediction. The same uniform prior distribution for $\boldsymbol{\theta}$ as in the centralized version (see Fig. 5.6a) is used.

The globally updated posterior distribution of $\boldsymbol{\theta}$ at time t_{100} is shown in Fig. 5.8. It has a peak near the true $\boldsymbol{\theta}$, which shows the correctness of the distributed algorithm. The predicted field compared with the true field at time t_{100} is shown in Fig. 5.9. Due to the construction of sub-regions, the interface areas between any of two sub-regions are not predicted. Notice that the prediction is not as good as in the 1-D scenario due to the effect of curse of dimensionality when we move from 1-D to 2-D spaces. The prediction quality can be improved by using more number of sensors at the cost of computational time. The running time of the distributed algorithm in this scenario is about several minutes due to the complexity of the 2-D problem under the same computational environment as the one used for the 1-D scenario. However, thanks to our proposed sequential sampling schemes, the running time does not grow with the number of measurements.

Chapter 6
New Efficient Spatial Model with Built-In Gaussian Markov Random Fields

Recently, there have been efforts to find a way to fit a computationally efficient Gaussian Markov Random Field (GMRF) on a discrete lattice to a Gaussian random field on a continuum space [86–88]. Such methods have been developed using a fitting with a weighted L_2-type distance [86], using a conditional-mean least-squares fitting [87], and for dealing with large data by fast Kriging [88]. It has been demonstrated that GMRFs with small neighborhoods can approximate Gaussian fields surprisingly well [86]. This approximated GMRF and its regression are very attractive for the resource-constrained mobile sensor networks due to its computational efficiency and scalability [89] as compared to the standard Gaussian process and its regression, which is not scalable as the number of observations increases.

Mobile sensing agents form an ad hoc wireless communication network in which each agent usually operates under a short communication range, with limited memory and computational power. For resource-constrained mobile sensor networks, developing distributed prediction algorithms for robotic sensors using only local information from local neighboring agents has been one of the most fundamental problems [42, 45, 46, 68, 112, 113].

In Sect. 6.1.1, a new class of Gaussian processes is proposed for resource-constrained mobile sensor networks. Such a Gaussian process builds on a GMRF [92] with respect to a proximity graph, e.g., the Delaunay graph of a set of vertices over a surveillance region. The formulas for predictive statistics are derived in Sect. 6.1.2. We propose a sequential prediction algorithm which is scalable to deal with sequentially sampled observations in Sect. 6.1.3. In Sect. 6.2, we develop a distributed and scalable statistical inference algorithm for a simple sampling scheme by applying the Jacobi over-relaxation and discrete-time average consensus algorithms. Simulation and experimental study demonstrate the usefulness of the proposed model and algorithms in Sect. 6.3.

Y. Xu et al., *Bayesian Prediction and Adaptive Sampling Algorithms for Mobile Sensor Networks*, SpringerBriefs in Control, Automation and Robotics, DOI 10.1007/978-3-319-21921-9_6

6.1 Spatial Prediction

In this section, we first propose a new class of Gaussian random fields with built-in Gaussian Markov Random Fields (GMRF) [92]. Then we show how to compute the prediction at any point of interest based on Gaussian process regression, and provide a sequential field prediction algorithm for mobile sensor networks.

6.1.1 Spatial Model Based on GMRF

Let $\gamma := (\gamma(\mathbf{p}_1), \cdots, \gamma(\mathbf{p}_m))^T \sim \mathbb{N}(0, \mathbf{Q}^{-1})$ be a zero-mean GMRF [92] with respect to an undirected graph $\mathcal{G} = (\mathcal{V}, \mathcal{E})$, where the location of vertex i is denoted by p_i in the surveillance region $\mathcal{Q}$. Such locations of vertices will be referred to as *generating points*. The inverse covariance matrix (precision matrix) $\mathbf{Q} \succ 0$ has the property $(\mathbf{Q})_{ij} \neq 0 \Leftrightarrow \{i, j\} \in \mathcal{E}$. If the graph $\mathcal{G}$ has small cardinalities of the neighbor sets, its precision matrix $\mathbf{Q}$ becomes sparse with many zeros in its entries. This plays a key role in computation efficiency of a GMRF which can be greatly exploited by the resource-constrained mobile sensor network.

The spatial field is modeled by a Gaussian process with a built-in GMRF defined as

$$z(\mathbf{s}) = \mu(\mathbf{s}) + \sum_{j=1}^{m} \lambda(\mathbf{s}, \mathbf{p}_j)\gamma(\mathbf{p}_j), \tag{6.1}$$

where $\lambda(\cdot, \cdot)$ is a weighting function. The new class of Gaussian processes is capable of representing a wide range of nonstationary Gaussian fields, by selecting

1. different number of generating points m,
2. different locations of generating points $\{\mathbf{p}_j \mid j = 1, \cdots, m\}$ over $\mathcal{Q}$,
3. a different structure of the precision matrix $\mathbf{Q}$, and
4. different weighting functions $\{\lambda(\cdot, \mathbf{p}_j) \mid j = 1, \cdots, m\}$.

Remark 6.1 The number of generating points could be determined by a model selection criterion such as the Akaike information criterion [114]. Similar to hyperparameter estimation in the standard Gaussian process regression, one can estimate all other parameters using maximum likelihood (ML) optimization [1, 53]. This is non-convex optimization and so the initial conditions need to be chosen carefully to avoid local minima. In our approach, we use basic structures for weighting functions and the precision matrix; however, we make them as functions of the locations of generating points. Different spatial resolutions can be obtained by a suitable choice of locations of generating points. As an example shown in Fig. 6.1, higher resolution can be obtained by higher density of generating points (see lower left corner). In this way, we only need to determine the locations of generating points. This approach will be demonstrated with real-world data in Sect. 6.3.4.

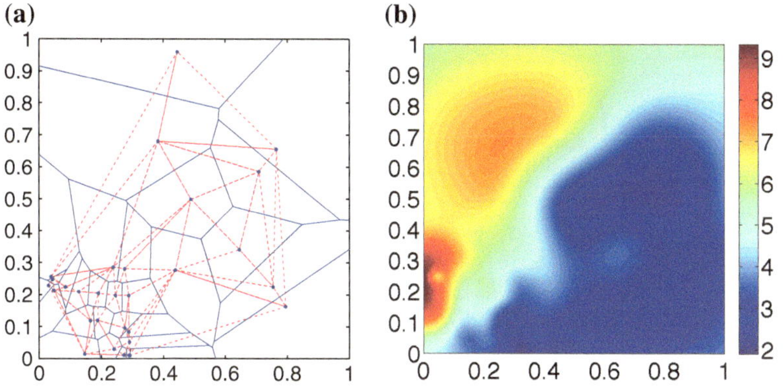

Fig. 6.1 **a** Generating points in *blue dots* and the associated Delaunay graph with edges in *red dotted lines*. The Voronoi partition is also shown in *blue solid lines*. **b** Gaussian random field with a built-in GMRF with respect to the Delaunay graph in (**a**)

6.1.2 Gaussian Process Regression

Suppose we have a collection of observations $\mathbf{y} := (y_1, \cdots, y_n)^T$ whose entries are sampled at the corresponding points $\mathbf{s}_1, \cdots, \mathbf{s}_n$. The noise corrupted measurement $y_i \in \mathbb{R}$ is given by

$$y_i = z(\mathbf{s}_i) + \epsilon_i,$$

where $\epsilon_i \overset{i.i.d.}{\sim} \mathbb{N}(0, \sigma_w^2)$ is an independent and identically distributed (i.i.d.) Gaussian white noise. We then have the following results.

Proposition 6.1 *Let* $\mathbf{\Lambda} \in \mathbb{R}^{n \times m}$ *be a matrix obtained by* $(\mathbf{\Lambda})_{ij} = \lambda(\mathbf{s}_i, \mathbf{p}_j)$ *and let* $\boldsymbol{\lambda} \in \mathbb{R}^m$ *be a vector obtained by* $(\boldsymbol{\lambda})_i = \lambda(\mathbf{s}_0, \mathbf{p}_i)$*, where* $\mathbf{s}_0$ *is a point of interest. Then the covariance matrix of* $\mathbf{y}$ *and the covariance between* $\mathbf{y}$ *and* $z(\mathbf{s}_0)$ *are given by*

$$\mathbf{C} := \mathbb{E}[(\mathbf{y} - \mathbb{E}\mathbf{y})(\mathbf{y} - \mathbb{E}\mathbf{y})^T] = \mathbf{\Lambda}\mathbf{Q}^{-1}\mathbf{\Lambda}^T + \sigma_w^2\mathbf{I},$$
$$\mathbf{k} := \mathbb{E}[(\mathbf{y} - \mathbb{E}\mathbf{y})z(\mathbf{s}_0)] = \mathbf{\Lambda}\mathbf{Q}^{-1}\boldsymbol{\lambda},$$

where $\mathbf{Q} \in \mathbb{R}^{m \times m}$ *is the precision matrix of the GMRF* $\gamma \in \mathbb{R}^m$.

Proof The (i, j)th element of the covariance matrix $\mathbf{C}$, i.e., the covariance between y_i and y_j, can be obtained by

$$\begin{aligned}
(\mathbf{C})_{ij} &= \mathrm{Cov}(z(\mathbf{s}_i), z(\mathbf{s}_j)) + \sigma_w^2 \delta_{ij} \\
&= \mathbb{E}(z(\mathbf{s}_i) - \mu(\mathbf{s}_i))(z(\mathbf{s}_j) - \mu(\mathbf{s}_j)) + \sigma_w^2 \delta_{ij} \\
&= \mathbb{E}\left(\sum_k \lambda(\mathbf{s}_i, \mathbf{p}_k)\gamma(\mathbf{p}_k)\right)\left(\sum_l \lambda(\mathbf{s}_j, \mathbf{p}_l)\gamma(\mathbf{p}_l)\right) + \sigma_w^2 \delta_{ij} \\
&= \mathbb{E}\left(\sum_{k,l} \lambda(\mathbf{s}_i, \mathbf{p}_k)\gamma(\mathbf{p}_k)\gamma(\mathbf{p}_l)\lambda(\mathbf{s}_j, \mathbf{p}_l)\right) + \sigma_w^2 \delta_{ij} \\
&= \sum_{k,l} \lambda(\mathbf{s}_i, \mathbf{p}_k)\mathbb{E}(\gamma(\mathbf{p}_k)\gamma(\mathbf{p}_l))\lambda(\mathbf{s}_j, \mathbf{p}_l) + \sigma_w^2 \delta_{ij} \\
&= \sum_{k,l} \lambda(\mathbf{s}_i, \mathbf{p}_k)(\mathbf{Q}^{-1})_{kl}\lambda(\mathbf{s}_j, \mathbf{p}_l) + \sigma_w^2 \delta_{ij}.
\end{aligned}$$

The ith element of the covariance vector $\mathbf{k}$, i.e., the covariance between y_i and $z(\mathbf{s}_0)$, can be obtained by

$$\begin{aligned}
(\mathbf{k})_i &= \mathrm{Cov}(z(\mathbf{s}_i), z(\mathbf{s}_0)) \\
&= \mathbb{E}(z(\mathbf{s}_i) - \mu(\mathbf{s}_i))(z(\mathbf{s}_0) - \mu(\mathbf{s}_0)) \\
&= \mathbb{E}\left(\sum_k \lambda(\mathbf{s}_i, \mathbf{p}_k)\gamma(\mathbf{p}_k)\right)\left(\sum_l \lambda(\mathbf{s}_0, \mathbf{p}_l)\gamma(\mathbf{p}_j)\right) \\
&= \mathbb{E}\left(\sum_{k,l} \lambda(\mathbf{s}_i, \mathbf{p}_k)\gamma(\mathbf{p}_k)\gamma(\mathbf{p}_l)\lambda(\mathbf{s}_0, \mathbf{p}_l)\right) \\
&= \sum_{k,l} \lambda(\mathbf{s}_i, \mathbf{p}_k)\mathbb{E}(\gamma(\mathbf{p}_k)\gamma(\mathbf{p}_l))\lambda(\mathbf{s}_0, \mathbf{p}_l) \\
&= \sum_{k,l} \lambda(\mathbf{s}_i, \mathbf{p}_k)(\mathbf{Q}^{-1})_{kl}\lambda(\mathbf{s}_0, \mathbf{p}_l),
\end{aligned}$$

whose matrix form completes the proof. □

By Proposition 6.1, we can make prediction at the point of interest s_0 using Gaussian process regression [53]. This is summarized by the following theorem.

Theorem 6.1 *For given y, the prediction of* $z_0 := z(\mathbf{s}_0)$ *at any location* $\mathbf{s}_0 \in \mathcal{Q}$ *is given by the conditional distribution*

$$z_0|\mathbf{y} \sim \mathbb{N}\left(\mu_{z_0|\mathbf{y}}, \sigma^2_{z_0|\mathbf{y}}\right),$$

where the predictive mean and variance are obtained by

$$\mu_{z_0|\mathbf{y}} = \mu(\mathbf{s}_0) + \boldsymbol{\lambda}^T \hat{\mathbf{Q}}^{-1} \hat{\mathbf{y}}, \tag{6.2}$$
$$\sigma^2_{z_0|\mathbf{y}} = \boldsymbol{\lambda}^T \hat{\mathbf{Q}}^{-1} \boldsymbol{\lambda},$$

with

$$\hat{\mathbf{Q}} = \mathbf{Q} + \sigma_w^{-2} \boldsymbol{\Lambda}^T \boldsymbol{\Lambda} \in \mathbb{R}^{m \times m},$$
$$\hat{\mathbf{y}} = \sigma_w^{-2} \boldsymbol{\Lambda}^T (\mathbf{y} - \boldsymbol{\mu}) \in \mathbb{R}^m.$$

Proof By using the Woodbury matrix identity (see Appendix A.2.1), the prediction mean can be obtained by

$$\begin{aligned}
\mu_{z_0|\mathbf{y}} &= \mu(\mathbf{s}_0) + \mathbf{k}^T \mathbf{C}^{-1} (\mathbf{y} - \boldsymbol{\mu}) \\
&= \mu(\mathbf{s}_0) + (\boldsymbol{\Lambda} \mathbf{Q}^{-1} \boldsymbol{\lambda})^T (\boldsymbol{\Lambda} \mathbf{Q}^{-1} \boldsymbol{\Lambda}^T + \sigma_w^2 \mathbf{I})^{-1} (\mathbf{y} - \boldsymbol{\mu}) \\
&= \mu(\mathbf{s}_0) + \boldsymbol{\lambda}^T \mathbf{Q}^{-1} \boldsymbol{\Lambda}^T (\boldsymbol{\Lambda} \mathbf{Q}^{-1} \boldsymbol{\Lambda}^T + \sigma_w^2 \mathbf{I})^{-1} (\mathbf{y} - \boldsymbol{\mu}) \\
&= \mu(\mathbf{s}_0) + \boldsymbol{\lambda}^T \mathbf{Q}^{-1} \boldsymbol{\Lambda}^T (\sigma_w^{-2} \mathbf{I} - \\
&\quad \sigma_w^{-2} \boldsymbol{\Lambda} (\mathbf{Q} + \sigma_w^{-2} \boldsymbol{\Lambda}^T \boldsymbol{\Lambda})^{-1} \boldsymbol{\Lambda}^T \sigma_w^{-2}) (\mathbf{y} - \boldsymbol{\mu}) \\
&= \mu(\mathbf{s}_0) + \boldsymbol{\lambda}^T (\sigma_w^{-2} \mathbf{Q}^{-1} - \\
&\quad \sigma_w^{-4} \mathbf{Q}^{-1} \boldsymbol{\Lambda}^T \boldsymbol{\Lambda} (\mathbf{Q} + \sigma_w^{-2} \boldsymbol{\Lambda}^T \boldsymbol{\Lambda})^{-1}) \boldsymbol{\Lambda}^T (\mathbf{y} - \boldsymbol{\mu}) \\
&= \mu(\mathbf{s}_0) + \boldsymbol{\lambda}^T \boldsymbol{\Xi} \boldsymbol{\Lambda}^T (\mathbf{y} - \boldsymbol{\mu}),
\end{aligned}$$

where

$$\begin{aligned}
\boldsymbol{\Xi} &= \sigma_w^{-2} \mathbf{Q}^{-1} - \sigma_w^{-4} \mathbf{Q}^{-1} \boldsymbol{\Lambda}^T \boldsymbol{\Lambda} (\mathbf{Q} + \sigma_w^{-2} \boldsymbol{\Lambda}^T \boldsymbol{\Lambda})^{-1} \\
&= \sigma_w^{-2} \mathbf{Q}^{-1} (\mathbf{Q} + \sigma_w^{-2} \boldsymbol{\Lambda}^T \boldsymbol{\Lambda}) (\mathbf{Q} + \sigma_w^{-2} \boldsymbol{\Lambda}^T \boldsymbol{\Lambda})^{-1} \\
&\quad - \sigma_w^{-4} \mathbf{Q}^{-1} \boldsymbol{\Lambda}^T \boldsymbol{\Lambda} (\mathbf{Q} + \sigma_w^{-2} \boldsymbol{\Lambda}^T \boldsymbol{\Lambda})^{-1} \\
&= (\sigma_w^{-2} \mathbf{I} + \sigma_w^{-4} \mathbf{Q}^{-1} \boldsymbol{\Lambda}^T \boldsymbol{\Lambda}) (\mathbf{Q} + \sigma_w^{-2} \boldsymbol{\Lambda}^T \boldsymbol{\Lambda})^{-1} \\
&\quad - \sigma_w^{-4} \mathbf{Q}^{-1} \boldsymbol{\Lambda}^T \boldsymbol{\Lambda} (\mathbf{Q} + \sigma_w^{-2} \boldsymbol{\Lambda}^T \boldsymbol{\Lambda})^{-1} \\
&= \sigma_w^{-2} (\mathbf{Q} + \sigma_w^{-2} \boldsymbol{\Lambda}^T \boldsymbol{\Lambda})^{-1}.
\end{aligned}$$

Similarly, the prediction error variance can be obtained by

$$\begin{aligned}
\sigma^2_{z_0|\mathbf{y}} &= \boldsymbol{\lambda}^T \mathbf{Q}^{-1} \boldsymbol{\lambda} - \mathbf{k}^T \mathbf{C}^{-1} \mathbf{k} \\
&= \boldsymbol{\lambda}^T \mathbf{Q}^{-1} \boldsymbol{\lambda} - (\boldsymbol{\Lambda} \mathbf{Q}^{-1} \boldsymbol{\lambda})^T (\boldsymbol{\Lambda} \mathbf{Q}^{-1} \boldsymbol{\Lambda}^T + \sigma_w^2 \mathbf{I})^{-1} (\boldsymbol{\Lambda} \mathbf{Q}^{-1} \boldsymbol{\lambda}) \\
&= \boldsymbol{\lambda}^T \left(\mathbf{Q}^{-1} - \mathbf{Q}^{-1} \boldsymbol{\Lambda}^T (\boldsymbol{\Lambda} \mathbf{Q}^{-1} \boldsymbol{\Lambda}^T + \sigma_w^2 \mathbf{I})^{-1} \boldsymbol{\Lambda} \mathbf{Q}^{-1} \right) \boldsymbol{\lambda} \\
&= \boldsymbol{\lambda}^T (\mathbf{Q} + \sigma_w^{-2} \boldsymbol{\Lambda}^T \boldsymbol{\Lambda})^{-1} \boldsymbol{\lambda},
\end{aligned}$$

where $\mathrm{Cov}(z(\mathbf{s}_0), z(\mathbf{s}_0)) = \boldsymbol{\lambda}^T \mathbf{Q}^{-1} \boldsymbol{\lambda}$ is obtained similarly as in Proposition 6.1. □

Remark 6.2 When the generating points $\{\mathbf{p}_1, \mathbf{p}_2, \cdots, \mathbf{p}_m\}$ are not known *a priori*, they can be estimated by maximizing the likelihood function. Given n observations $\mathbf{y} = (y_1, y_2, \cdots, y_n)^T$ sampled at $\{\mathbf{s}_1, \mathbf{s}_2, \cdots, \mathbf{s}_n\}$, the log likelihood of $\mathbf{y}$ is given by

$$\log \pi(\mathbf{y}) = -\frac{1}{2}(\mathbf{y} - \mu)^T \mathbf{C}^{-1}(\mathbf{y} - \mu) - \frac{1}{2} \log \det \mathbf{C} - \frac{n}{2} \log 2\pi,$$

where $\mathbf{C} = \boldsymbol{\Lambda} \mathbf{Q}^{-1} \boldsymbol{\Lambda}^T + \sigma_w^2 \mathbf{I}$ is the covariance matrix of $\mathbf{y}$. the maximum likelihood estimate of the generating points can be obtained via solving the following optimization problem.

$$\hat{\mathbf{p}}_{\mathrm{ML}} = \arg \max_{\mathbf{p}} \log \pi(\mathbf{y}). \tag{6.3}$$

Remark 6.3 Note that the number of generating points m is fixed and the number of observations n may grow in time, and so in general we consider $m \ll n$. Theorem 6.1 shows that only the inversion of an $m \times m$ matrix $\hat{\mathbf{Q}} = \mathbf{Q} + \sigma_w^{-2} \boldsymbol{\Lambda}^T \boldsymbol{\Lambda}$ is required in order to compute the predictive distribution of the field at any point. The computational complexity grows linearly with the number of observations, i.e., $O(nm^2)$, compare to the standard Gaussian process regression which requires $O(n^3)$. Moreover, it enables a scalable prediction algorithm for sequential measurements.

In what follows, we present a sequential field prediction algorithm for sequential observations by exploiting the results of Theorem 6.1.

6.1.3 Sequential Prediction Algorithm

Consider a sensor network consisting of N mobile sensing agents distributed in the surveillance region $\mathcal{Q}$. The index of the robotic sensors is denoted by $\mathcal{I} := \{1, \cdots, N\}$. The sensing agents sample the environmental field at time $t \in \mathbb{Z}_{>0}$ and send the observations to a central station which is in charge of the data fusion.

At time t, agent i makes an observation $y_i(t)$ at location $\mathbf{s}_i(t)$. Denote the collection of observations at time t by $\mathbf{y}_t := (y_1(t), \cdots, y_N(t))^T$. We have the following proposition.

Proposition 6.2 *At time $t \in \mathbb{Z}_{>0}$, the predictive mean and variance at any point of interest can be obtained via* (6.2) *with*

$$\begin{aligned} \hat{\mathbf{Q}}_t &= \hat{\mathbf{Q}}_{t-1} + \sigma_w^{-2} \boldsymbol{\Lambda}_t^T \boldsymbol{\Lambda}_t, \quad \hat{\mathbf{Q}}_0 = \mathbf{Q} \\ \hat{\mathbf{y}}_t &= \hat{\mathbf{y}}_{t-1} + \sigma_w^{-2} \boldsymbol{\Lambda}_t^T (\mathbf{y}_t - \boldsymbol{\mu}_t), \quad \hat{\mathbf{y}}_0 = \mathbf{0}, \end{aligned}$$

where $(\boldsymbol{\Lambda}_t)_{ij} = \lambda(\mathbf{s}_i(t), \mathbf{s}_j(t))$, and $(\boldsymbol{\mu}_t)_i = \mu(\mathbf{s}_i(t))$.

Table 6.1 Sequential algorithm for field prediction

Input:
a set of target points $\mathcal{S}$
Output:
(1) prediction mean $\{\hat{z}(\mathbf{s}_0) \mid \mathbf{s}_0 \in \mathcal{S}\}$
(2) prediction error variance $\{\sigma^2(\mathbf{s}_0) \mid \mathbf{s}_0 \in \mathcal{S}\}$
Assumption:
(1) the central station knows $\mathbf{p}$, $\mathbf{Q}$, and $\lambda(\cdot,\cdot)$
(2) the central station initially has $\hat{\mathbf{Q}} \leftarrow \mathbf{Q}$, $\hat{\mathbf{y}} \leftarrow 0$

At time t, agent $i \in \mathcal{I}$ in the network does:

```
take measurement y_i from its current location s_i
send the measurement (s_i, y_i) to the central station
```

At time t, the central station does:

- obtain measurements $\{(\mathbf{s}_\ell, y_\ell) \mid \forall \ell \in \mathcal{I}\}$ from mobile sensors
- compute $\mathbf{\Lambda}$ via $(\mathbf{\Lambda})_{ij} = \lambda(\mathbf{s}_i, \mathbf{p}_j)$
- update $\hat{\mathbf{Q}} \leftarrow \hat{\mathbf{Q}} + \sigma_w^{-2}\mathbf{\Lambda}^T\mathbf{\Lambda}$
- update $\hat{\mathbf{y}} \leftarrow \hat{\mathbf{y}} + \sigma_w^{-2}\mathbf{\Lambda}^T(\mathbf{y} - \boldsymbol{\mu})$, where $(\boldsymbol{\mu})_i = \mu(\mathbf{s}_i)$
- **for** $\mathbf{s}_0 \in \mathcal{S}$ **do**
  - compute $(\boldsymbol{\lambda})_i$ via $\lambda(\mathbf{s}_0, \mathbf{p}_i)$
  - compute $\hat{z}(\mathbf{s}_0) = \mu(\mathbf{s}_0) + \boldsymbol{\lambda}^T\hat{\mathbf{Q}}^{-1}\hat{\mathbf{y}}$
  - compute $\sigma^2(\mathbf{s}_0) = \boldsymbol{\lambda}^T\hat{\mathbf{Q}}^{-1}\boldsymbol{\lambda}$
- **end for**

Proof The result can be obtained easily by noting that $\mathbf{A}^T\mathbf{A} = \mathbf{A}_1^T\mathbf{A}_1 + \mathbf{A}_2^T\mathbf{A}_2$, where $\mathbf{A} = (\mathbf{A}_1^T, \mathbf{A}_2^T)^T$. □

Based on Proposition 6.2, we present a sequential field prediction algorithm using mobile sensor networks in Table 6.1.

6.2 Distributed Spatial Prediction

In this section, we propose a distributed approach, in which robotic sensors exchange only local information between neighbors, to implement the field prediction effectively fusing all observations collected by all sensors correctly. This distributed approach can be implemented for a class of weighting functions $\lambda(\cdot,\cdot)$ in (6.1) that have compact supports. In particular, we consider the weighting function defined by

$$\lambda(\mathbf{s}, \mathbf{p}_j) = \lambda(\|\mathbf{s} - \mathbf{p}_j\|/r), \tag{6.4}$$

where

$$\lambda(h) := \begin{cases} (1-h)\cos(\pi h) + \frac{1}{\pi}\sin(\pi h), & h \le 1, \\ 0, & \text{otherwise.} \end{cases}$$

Notice that the weighting function $\lambda(\cdot,\cdot)$ in (6.4) has a compact support, i.e., $\lambda(\mathbf{s}, \mathbf{p}_j)$ is nonzero if and only if the distance $\|\mathbf{s} - \mathbf{p}_j\|$ is less than the support $r \in \mathbb{R}_{>0}$.

6.2.1 Distributed Computation

We first briefly introduce distributed algorithms for solving linear systems and computing the averages. They will be used as major tools for distributed implementation of field prediction.

- **Jacobi over-relaxation method:** The Jacobi over-relaxation (JOR) [112] method provides an iterative solution of a linear system $\mathbf{A}\mathbf{x} = \mathbf{b}$, where $\mathbf{A} \in \mathbb{R}^{n\times n}$ is a nonsingular matrix and $\mathbf{x}, \mathbf{b} \in \mathbb{R}^n$. If agent i knows the $\text{row}_i(\mathbf{A}) \in \mathbb{R}^n$ and b_i, and $a_{ij} = (\mathbf{A})_{ij} = 0$ if agent i and agent j are not neighbors, then the recursion is given by

$$x_i^{(k+1)} = (1-h)x_i^{(k)} + \frac{h}{a_{ii}}\left(b_i - \sum_{j\in\mathcal{N}_i} a_{ij}x_j^{(k)}\right). \tag{6.5}$$

  This JOR algorithm converges to the solution of $Ax = b$ from any initial condition if $h < 2/n$ [45]. At the end of the algorithm, agent i knows the ith element of $\mathbf{x} = \mathbf{A}^{-1}\mathbf{b}$.
- **Discrete-time average consensus:** The Discrete-time average consensus (DAC) provides a way to compute the arithmetic mean of elements in the a vector $\mathbf{c} \in \mathbb{R}^n$. Assume the graph is connected. If agent i knows the ith element of $\mathbf{c}$, the network can compute the arithmetic mean via the following recursion [113]

$$x_i^{(k+1)} = x_i^{(k)} + \epsilon \sum_{j\in\mathcal{N}_i} a_{ij}(x_j^{(k)} - x_i^{(k)}), \tag{6.6}$$

  with initial condition $\mathbf{x}(0) = \mathbf{c}$, where $a_{ij} = 1$ if $j \in \mathcal{N}_i$ and 0 otherwise, $0 < \epsilon < 1/\Delta$, and $\Delta = \max_i(\sum_{j\neq i} a_{ij})$ is the maximum degree of the network. After the algorithm converges, all node in the network know the average of $\mathbf{c}$, i.e., $\sum_{i=1}^{n} c_i/n$.

6.2.2 Distributed Prediction Algorithm

Consider a GMRF with respect to a proximity graph $\mathcal{G} = (\mathcal{V}, \mathcal{E})$ that generates a Gaussian random field in (6.1). The index of the generating points is denoted by $\mathcal{V} := \{1, \cdots, n\}$. The location of the ith generating point is $\mathbf{p}_i$. The edges of the graph are considered to be $\mathcal{E} := \left\{\{i, j\} \mid \|\mathbf{p}_i - \mathbf{p}_j\| \le R\right\}$, where R is a constant that ensures the graph is connected.

Consider a mobile sensor network consisting of N mobile sensing agents distributed in the surveillance region $\mathcal{Q}$. For simplicity, we assume that the number of agents is equal to the number of generating points, i.e., $N = m$. The index of the robotic sensors is denoted by $\mathcal{I} := \{1, \cdots, m\}$. The location of agent i is denoted by $\mathbf{s}_i$.

The assumptions made for the resource-constrained mobile sensor networks are listed as follows.

A.1 Agent i is in charge of sampling at point $\mathbf{s}_i$ within a r-disk centered at $\mathbf{p}_i$, i.e., $\|\mathbf{s}_i - \mathbf{p}_i\| < r$.
A.2 r is the radius of the support of the weighting function in (6.4) and also satisfies that $0 < r < \frac{R}{2}$.
A.3 Agent i can only locally communicate with neighbors in $\mathcal{N}_i := \{j \in \mathcal{I} \mid \{i, j\} \in \mathcal{E}\}$ defined by the connected proximity graph $\mathcal{G} = (\mathcal{V}, \mathcal{E})$.
A.4 Agent i knows $\text{row}_i(\mathbf{Q})$, i.e., the ith row of $\mathbf{Q}$, where $(\mathbf{Q})_{ij} \neq 0$ if and only if $j \in \{i\} \cup \mathcal{N}_i$.

Remark 6.4 As in A.1, it is reasonable to have at least one agent collect measurements that are correlated with a random variable from a single generating point. This sampling rule may be modified such that a single agent dynamically samples for multiple generating points or more number of agents samples for a generating point depending on available resources. Since there is at least one agent in charge of a generating point by A.1, it is natural to have A.3 and A.4 taking advantage of the proximity graph for the GMRF. Notice that each agent only knows local information of $\mathbf{Q}$ as described in A.4.

An illustration of agent ℓ sampling a measurement at point $\mathbf{s}_\ell$ in the intersection of the supports of the weighting functions of $\mathbf{p}_i$ and $\mathbf{p}_j$ is shown in Fig. 6.2.

From A.1 and A.2, since $R > 2r$, we have $\lambda(\mathbf{s}_\ell, \mathbf{p}_i) = 0$ if $\ell \notin \mathcal{N}_i$. Thus the matrix $\hat{\mathbf{Q}} = \mathbf{Q} + \sigma_w^{-2}\mathbf{\Lambda}^T\mathbf{\Lambda} \in \mathbb{R}^{m\times m}$ and the vector $\hat{\mathbf{y}} = \sigma_w^{-2}\mathbf{\Lambda}^T(\mathbf{y} - \boldsymbol{\mu}) \in \mathbb{R}^m$ can be obtained in the following form.

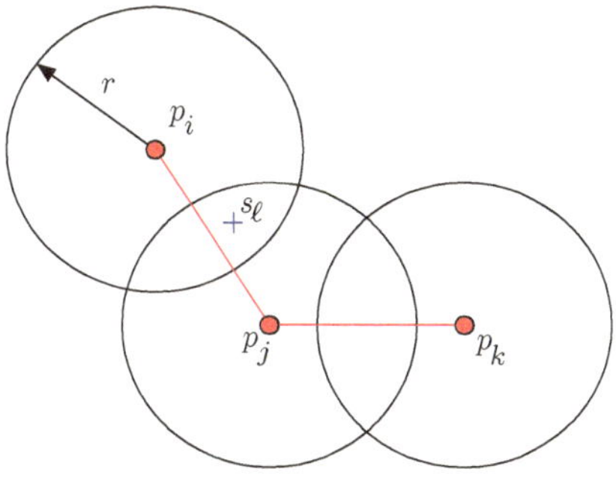

Fig. 6.2 Example of computing $(\mathbf{\Lambda}^T\mathbf{\Lambda})_{ij} = \lambda(\mathbf{s}_\ell, \mathbf{p}_i)\lambda(\mathbf{s}_\ell, \mathbf{p}_j)$

$$
\begin{aligned}
(\hat{\mathbf{Q}})_{ij} &= (\mathbf{Q})_{ij} + \sigma_w^{-2} \sum_{\ell \in \{\{i\} \cup \mathcal{N}_i\} \cap \{\{j\} \cup \mathcal{N}_j\}} \lambda(\mathbf{s}_\ell, \mathbf{p}_i)\lambda(\mathbf{s}_\ell, \mathbf{p}_j), \\
(\hat{\mathbf{y}})_i &= \sigma_w^{-2} \sum_{\ell \in \{i\} \cup \mathcal{N}_i} \lambda(\mathbf{s}_\ell, \mathbf{p}_i)(y_\ell - \mu_\ell).
\end{aligned}
\tag{6.7}
$$

Notice that $\hat{\mathbf{Q}}$ has the same sparsity as $\mathbf{Q}$. From (6.7), A.3 and A.4, agent i can compute $\text{row}_i(\hat{\mathbf{Q}})$ and $(\hat{\mathbf{y}})_i$ by using only local information from neighbors. Using $\text{row}_i(\hat{\mathbf{Q}})$ and $(\boldsymbol{\lambda})_i$, agent i can obtain the ith element in the vector $\hat{\mathbf{Q}}^{-1}\lambda = (\mathbf{Q} + \sigma_w^{-2}\boldsymbol{\Lambda}^T\boldsymbol{\Lambda})^{-1}\boldsymbol{\lambda}$ via JOR by using only local information. Finally, using $(\hat{\mathbf{y}})_i$ and $(\boldsymbol{\lambda})_i$ the prediction mean and variance can be obtained via the discrete-time average consensus algorithm. Notice that the sequential update of $\hat{\mathbf{Q}}$ and $\hat{\mathbf{y}}$ for sequential observations proposed in Sect. 6.1.3 can be also applied to the distributed algorithm. The distributed algorithm for sequential field prediction under assumptions A.1–4 is summarized in Table 6.2.

Table 6.2 Distributed algorithm for sequential field prediction

Input:
(1) a set of target points $\mathcal{S}$
(2) the topology of sensor network $\mathcal{G} = (\mathcal{I}, \mathcal{E})$ in which $\mathcal{E} := \{\{i,j\} \mid \|\mathbf{p}_i - \mathbf{p}_j\| \le R\}$
Output:
(1) prediction mean $\{\mu_{z_0|\mathbf{y}} \mid \mathbf{s}_0 \in \mathcal{S}\}$
(2) prediction error variance $\left\{\sigma^2_{z_0|\mathbf{y}} \mid \mathbf{s}_0 \in \mathcal{S}\right\}$
Assumption:
(A1) agent $i \in \mathcal{I}$ is in charge of sampling at point $\mathbf{s}_i$ within a r-disk centered at $\mathbf{p}_i$, *i.e.*, $\|\mathbf{s}_i - \mathbf{p}_i\| < r$
(A2) the radius of the support of the weighting function satisfies $0 < r < \frac{R}{2}$
(A3) agent $i \in \mathcal{I}$ can only locally communicate with neighbors $\mathcal{N}_i := \{j \in \mathcal{I} \mid \{i,j\} \in \mathcal{E}\}$ defined by the connected graph $\mathcal{G} = (\mathcal{V}, \mathcal{E})$
(A4) agent $i \in \mathcal{I}$ initially has $\text{row}_i(\hat{\mathbf{Q}}) \leftarrow \text{row}_i(\mathbf{Q})$, $(\hat{\mathbf{y}})_i \leftarrow 0$

At time t, agent $i \in \mathcal{I}$ in the network does the following concurrently:
1: take measurement y_i from its current location $\mathbf{s}_i$
2: update $\text{row}_i(\hat{\mathbf{Q}}) \leftarrow \text{row}_i(\hat{\mathbf{Q}}) + \text{row}_i(\sigma_w^{-2}\boldsymbol{\Lambda}^T\boldsymbol{\Lambda})$ by exchanging information from neighbors $\mathcal{N}_i$
3: update $(\hat{\mathbf{y}})_i \leftarrow (\hat{\mathbf{y}})_i + (\sigma_w^{-2}\boldsymbol{\Lambda}^T(\mathbf{y} - \boldsymbol{\mu}))_i$ by exchanging information from neighbors $\mathcal{N}_i$
4: **for** $\mathbf{s}_0 \in \mathcal{S}$ **do**
5: compute $(\boldsymbol{\lambda})_i = \lambda(\mathbf{s}_0, \mathbf{p}_i)$
6: compute $(\hat{\mathbf{Q}}^{-1}\boldsymbol{\lambda})_i$ via JOR
7: compute $\mu_{z_0|\mathbf{y}} = \mu(\mathbf{s}_0) + \boldsymbol{\lambda}^T\hat{\mathbf{Q}}^{-1}\hat{\mathbf{y}}$ via DAC
8: compute $\sigma^2_{z_0|\mathbf{y}} = \boldsymbol{\lambda}^T\hat{\mathbf{Q}}^{-1}\boldsymbol{\lambda}$ via DAC
9: **end for**

The number of robotic sensors and the sampling rule can be modified or optimized to maintain a better quality of the prediction and the corresponding distributed algorithm may be derived in a same way accordingly.

6.3 Simulation and Experiment

In this section, we apply the proposed schemes to both simulation and experimental study.

6.3.1 Simulation

We first apply our proposed prediction algorithms to a numerically generated Gaussian random field $z(\cdot)$ based on a GMRF with respect to a graph $\mathcal{G} = (\mathcal{V}, \mathcal{E})$ defined in (6.1). The mean function $\mu(\cdot)$ is assumed to be constant and $\mu = 5$ is used in the simulation. We assume the generating points of the GMRF, indexed by $\mathcal{V} = \{1, \cdots, n\}$ where $n = 30$, are located at $\{\mathbf{p}_1, \cdots, \mathbf{p}_n\}$ in a 2-D unit area $\mathcal{Q}$. The edges of the graph are assumed to be $\mathcal{E} := \{\{i, j\} \mid \|\mathbf{p}_i - \mathbf{p}_j\| \leq R\}$, where $R = 0.4$.

The GMRF $\gamma = (\gamma(\mathbf{p}_1), \cdots, \gamma(\mathbf{p}_n))^T$ has a zero-mean and the precision matrix $\mathbf{Q}$ is given by

$$(\mathbf{Q})_{ij} = \begin{cases} |\mathcal{N}(i)| + c_0, & \text{if } j = i, \\ -1, & \text{if } j \in \mathcal{N}(i), \\ 0, & \text{otherwise,} \end{cases}$$

where $|\mathcal{N}(i)|$ denotes the degree of node i, i.e., the number of connections it has to other nodes, $c_0 = 0.1$ is used to ensure $\mathbf{Q}$ is positive definite since a Hermitian diagonally dominant matrix with real non-negative diagonal entries is positive semi-definite [92]. We use compactly supported weighting functions defined in (6.4) for both centralized and distributed schemes with different support r. The sensor noise level is given by $\sigma_w = 0.5$. Since the optimal sampling is beyond the scope of this chapter, in the simulation, we use a random sampling strategy in which robotic sensors sample at random locations at each time instance.

6.3.2 Centralized Scheme

We first consider a scenario in which $N = 5$ agents take samples in the surveillance region $\mathcal{D}$ at certain time instance $t \in \mathbb{Z}_{>0}$ and send the observations to a central station in which the prediction of the field is made.

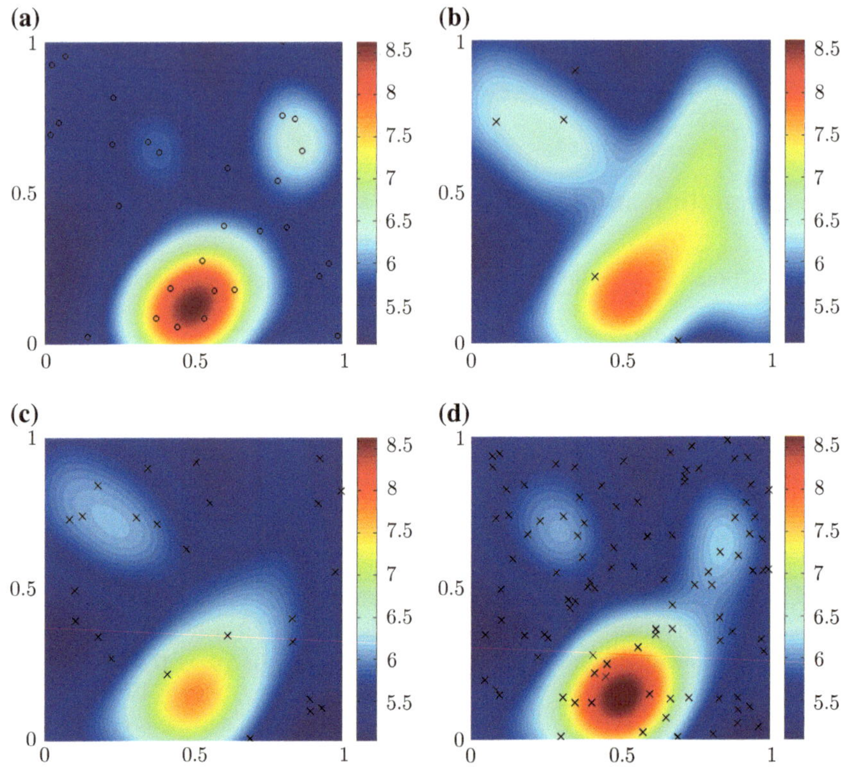

Fig. 6.3 Simulation results for the centralized scheme. **a** The true field, **b** the predicted field at time $t = 1$, **c** the predicted field at time $t = 5$, **d** the predicted field at time $t = 20$. The generating points are shown in *black circles*, and the sampling locations are shown in *black crosses*

The Gaussian random field $z(\cdot)$ is shown in Fig. 6.3a with the $n = 30$ generating points of the built-in GMRF shown in black circles. The predicted field at times $t = 1, t = 5$, and $t = 20$ are shown in Figs. 6.3b, c and d, respectively. The sampling locations are shown in black crosses. Clearly, the predicted field gets closer to the true field as the number of observations increases. The computational time for field prediction at each time instance remains fixed due to the nice structure of the proposed Gaussian field in (6.1) and its consequent results from Theorem 6.1.

6.3.3 Distributed Scheme

Next, we consider a scenario in which prediction is implemented in a distributed fashion (Table 6.2) under assumptions A.1–4 for the resource-constrained mobile sensor network in Sect. 6.2.2. In particular, $N = 30$ robotic sensors are distributed

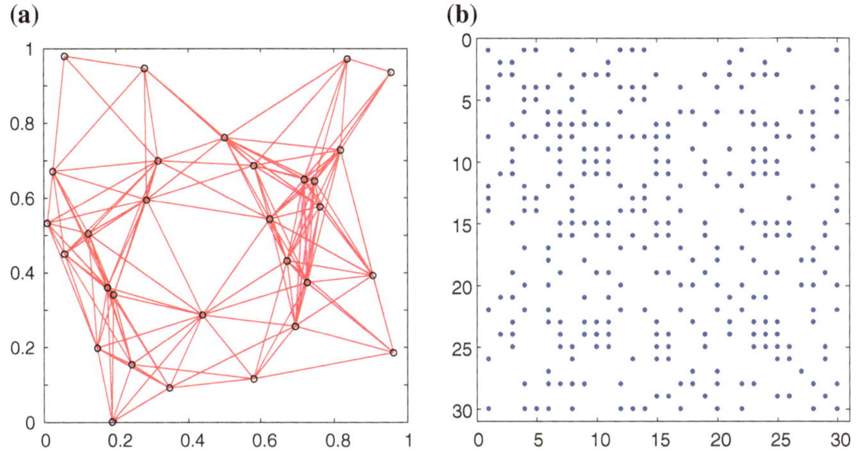

Fig. 6.4 **a** Graph $\mathcal{G} = (\mathcal{V}, \mathcal{E})$. **b** Sparsity structure of the precision matrix $\mathbf{Q}$

according to the graph $\mathcal{G} = (\mathcal{V}, \mathcal{E})$, which is connected. Agent i is in charge of the sampling with in a r-disk centered at p_i, where the support $r = 0.2$ is used. Agent i has a fixed neighborhood, i.e., $\mathcal{N}(i) = \{j \mid \{i, j\} \in \mathcal{E}\}$. In the simulation, $h = 0.02$ in (6.5) and $\epsilon = 0.02$ in (6.6) are chosen to ensure the convergence of the JOR algorithm and the DAC algorithm.

Figure 6.4a shows the underlying graph $\mathcal{G} = (\mathcal{V}, \mathcal{E})$ for the GMRF with the generating points denoted by black circles and the edges in red lines. The sparsity of the precision matrix $\mathbf{Q}$ is shown in Fig. 6.4b. Notice that only 316 out of 900 elements in $\mathbf{Q}$ are nonzero which enables the efficient distributed computation. The true and the predicted fields at time $t = 5$ are shown in Figs. 6.5a, b, respectively. The normalized RMS error computed over about 10000 grid points at time $t = 5$ is 7.8 %. The computational time at each time instance remains fixed due to the nice structure of the proposed Gaussian field in (6.1) and its consequent results from Theorem 6.1.

6.3.4 Experiment

In order to show the practical usefulness of the proposed approach, we apply the centralized scheme in Theorem 6.1 on an experimentally obtained observations. We first measured depth values of a terrain on grid points by using a Microsoft Kinect sensor [115] as shown in Fig. 6.6a. As pointed out in Remark 6.1, we make the structures of weighting functions and the precision matrix as functions of the locations of generating points. In particular, two generating points are neighbors if and only if their corresponding Voronoi cells intersect. The individual weighting function takes the same form as in (6.4) and its support size r_i is selected to be the largest distance between the generating point i and it's neighbors. We then predict the

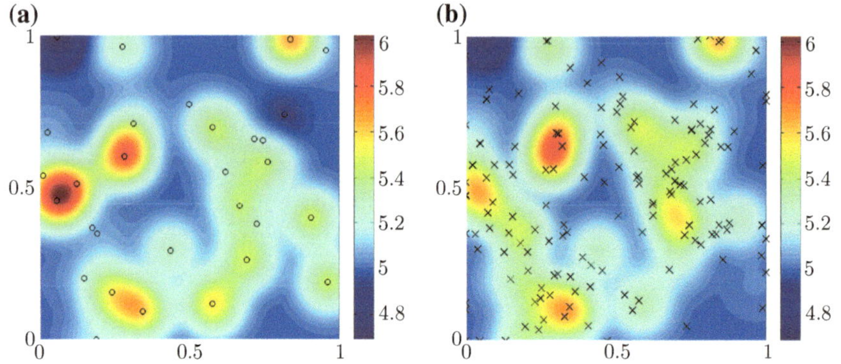

Fig. 6.5 Simulation results for the distributed scheme. **a** The true field, **b** the predicted field at time $t = 5$. The generating points are shown in *circles*, and the sampling locations are shown in crosses

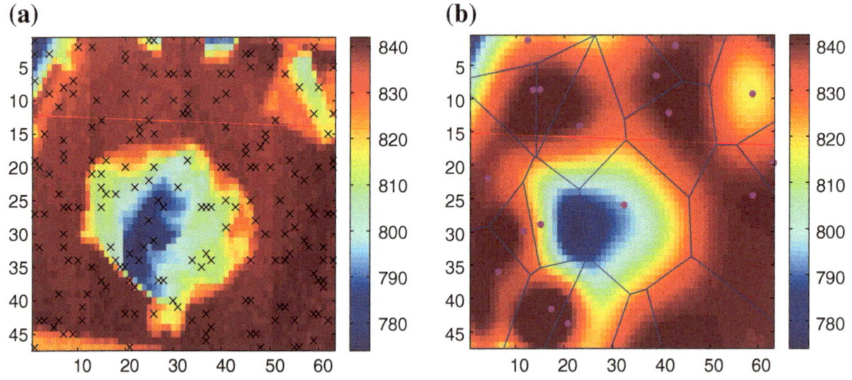

Fig. 6.6 **a** True field on grid positions obtained by the Kinect sensor and randomly sampled positions indicated in black crosses. **b** The fitted Gaussian random field with a build-in GMRF with respect to the Delaunay graph

field by our model with 20 estimated generating points given by the ML estimator in (6.3) using a subset of experimental observations, i.e., 200 randomly sampled observations denoted by crosses in Fig. 6.6a. The estimated positions of generating points along with the predicted field are shown in Fig. 6.6b. In this experiment, it is clear to see that our approach effectively produces the predicted field, which is very close to the true field for the case of unknown generating points.

Chapter 7
Fully Bayesian Spatial Prediction Using Gaussian Markov Random Fields

In this chapter, we consider the problem of predicting a large scale spatial field using successive noisy measurements obtained by mobile sensing agents. The physical spatial field of interest is discretized and modeled by a Gaussian Markov random field (GMRF) with unknown hyperparameters. From a Bayesian perspective, we design a sequential prediction algorithm to exactly compute the predictive inference of the random field. The main advantages of the proposed algorithm are: (1) the computational efficiency due to the sparse structure of the precision matrix, and (2) the scalability as the number of measurements increases. Thus, the prediction algorithm correctly takes into account the uncertainty in hyperparameters in a Bayesian way and also is scalable to be usable for the mobile sensor networks with limited resources. An adaptive sampling strategy is also designed for mobile sensing agents to find the most informative locations in taking future measurements in order to minimize the prediction error and the uncertainty in hyperparameters. The effectiveness of the proposed algorithms is illustrated by a numerical experiment.

In Chap. 5, we designed a sequential Bayesian prediction algorithm to deal with unknown bandwidths by using a compactly supported kernel and selecting a subset of collected measurements. In this chapter, we instead seek a fully Bayesian approach over a discretized surveillance region such that the Bayesian spatial prediction utilizes all collected measurements in a scalable fashion. In contrast to this chapter, Chap. 6 focuses on efficient spatial modeling using a GMRF with known hyperparameters. A distributed version of the prediction algorithm in this chapter for a special case can be found in [5].

In Sect. 7.1, we model the physical spatial field as a GMRF with unknown hyperparameters and formulate the estimation problem from a Bayesian point of view. In Sect. 7.2, we design a sequential Bayesian estimation algorithm to effectively and efficiently compute the exact predictive inference of the spatial field. The proposed

Y. Xu et al., *Bayesian Prediction and Adaptive Sampling Algorithms for Mobile Sensor Networks*, SpringerBriefs in Control, Automation and Robotics, DOI 10.1007/978-3-319-21921-9_7

algorithm often takes only seconds to run even for a very large spatial field, as will be demonstrated in this chapter. Moreover, the algorithm is scalable in the sense that the running time does not grow as the number of observations increases. In particular, the scalable prediction algorithm does not rely on the subset of samples to obtain scalability (as was done in Chap. 5), correctly fusing all collected measurements. In Sect. 7.4, an adaptive sampling strategy for mobile sensor networks is designed to largely improve the quality of prediction and to reduce the uncertainty in the hyperparameter estimation simultaneously. We demonstrate the effectiveness through a simulation study in Sect. 7.5.

7.1 Spatial Field Model

In what follows, we specify the models for the spatial field and the mobile sensor network. Notice that in this chapter, we slightly change notation for notational simplicity.

Let $\mathcal{Q}_* \subset \mathbb{R}^D$ denote the spatial field of interest. We discretize the field into n_* spatial sites $\mathcal{S}_* := \{\mathbf{s}_1, \ldots, \mathbf{s}_{n_*}\}$ and let $\mathbf{z}_* = (z_1, \ldots, z_{n_*})^T \in \mathbb{R}^{n_*}$ be the value of the field (e.g., the temperature). Due to the irregular shape a spatial field may have, we extend the field such that $n \geq n_*$ sites denoted by $\mathcal{S} := \{\mathbf{s}_1, \ldots, \mathbf{s}_n\}$ are on a regular grid. The latent variable $z_i := z(\mathbf{s}_i) \in \mathbb{R}$ is modeled by

$$z_i = \mu(\mathbf{s}_i) + \eta_i, \quad \forall 1 \leq i \leq n, \tag{7.1}$$

where $\mathbf{s}_i \in \mathcal{S} \subset \mathbb{R}^D$ is the ith site location. The mean function $\mu : \mathbb{R}^D \to \mathbb{R}$ is defined as

$$\mu(\mathbf{s}_i) = f(\mathbf{s}_i)^T \beta,$$

where $f(\mathbf{s}_i) = (f_1(\mathbf{s}_i), \ldots, f_p(\mathbf{s}_i))^T \in \mathbb{R}^p$ is a known regression function, and $\beta = (\beta_1, \ldots, \beta_p)^T \in \mathbb{R}^p$ is an unknown vector of regression coefficients. We define $\eta = (\eta_1, \ldots, \eta_n)^T \in \mathbb{R}^n$ as a zero-mean Gaussian Markov random field (GMRF) [92] denoted by

$$\eta \sim \mathrm{N}\left(\mathbf{0}, \mathbf{Q}_{\eta|\theta}^{-1}\right),$$

where the inverse covariance matrix (or precision matrix) $\mathbf{Q}_{\eta|\theta} \in \mathbb{R}^{n \times n}$ is a function of a hyperparameter vector $\theta \in \mathbb{R}^M$.

There exists many different choices of the GMRF (i.e., the precision matrix $\mathbf{Q}_{\eta|\theta}$) [92]. For instance, we can choose one with the full conditionals in (7.2) (with obvious notation as shown in [92]).

$$
\begin{aligned}
\mathbb{E}(\eta_i|\eta_{-i},\boldsymbol{\theta}) &= \frac{1}{4+a^2}\left(2a \begin{array}{ccccc} \circ&\circ&\circ&\circ&\circ \\ \circ&\circ&\bullet&\circ&\circ \\ \circ&\bullet&\circ&\bullet&\circ \\ \circ&\circ&\bullet&\circ&\circ \\ \circ&\circ&\circ&\circ&\circ \end{array} - 2 \begin{array}{ccccc} \circ&\circ&\circ&\circ&\circ \\ \circ&\bullet&\circ&\bullet&\circ \\ \circ&\circ&\circ&\circ&\circ \\ \circ&\bullet&\circ&\bullet&\circ \\ \circ&\circ&\circ&\circ&\circ \end{array} - 1 \begin{array}{ccccc} \circ&\circ&\bullet&\circ&\circ \\ \circ&\circ&\circ&\circ&\circ \\ \bullet&\circ&\circ&\circ&\bullet \\ \circ&\circ&\circ&\circ&\circ \\ \circ&\circ&\bullet&\circ&\circ \end{array} \right), \\
\mathrm{Var}(\eta_i|\eta_{-i},\boldsymbol{\theta}) &= (4+a^2)\kappa.
\end{aligned}
\tag{7.2}
$$

Figure 7.1 displays the elements of the precision matrix related to a single location that explains (7.2). The hyperparameter vector is defined as $\boldsymbol{\theta} = (\kappa, \alpha)^T \in \mathbb{R}^2_{>0}$, where $\alpha = a - 4$. The resulting GMRF accurately represents a Gaussian random field with the Matérn covariance function [116]

$$
C(r) = \sigma_f^2 \frac{2^{1-\nu}}{\Gamma(\nu)} \left(\frac{\sqrt{2\nu} r}{\ell} \right)^{\nu} K_{\nu}\left(\frac{\sqrt{2\nu} r}{\ell} \right),
$$

where $K_\nu(\cdot)$ is a modified Bessel function [53], with order $\nu = 1$, a bandwidth $\ell = 1/\sqrt{\alpha}$, and vertical scale $\sigma_f^2 = 1/4\pi\alpha\kappa$. The hyperparameter $\alpha > 0$ guarantees the positive definiteness of the precision matrix $\mathbf{Q}_{\eta|\theta}$. In the case where $\alpha = 0$, the resulting GMRF is a second-order polynomial intrinsic GMRF [92, 117]. Notice that the precision matrix is sparse which contains only small number of non-zero elements. This property will be exploited for fast computation in the following sections.

Example 7.1 Consider a spatial field of interest $\mathcal{Q}_* \in [0, 100] \times [0, 50]$. We first divide the spatial field into a 100×50 regular grid with equal areas 1, which makes $n_* = 5000$. We then extend the the field such that 120×70 grids (i.e., $n = 8400$) are constructed on the extended field $\mathcal{Q} = [-10, 110] \times [-10, 60]$. The precision matrix $\mathbf{Q}_{\eta|\theta}$ introduced above is chosen with the regular lattices wrapped on a torus [92]. In this case, only 0.15 % elements in the sparse matrix $\mathbf{Q}_{\eta|\theta}$ are non-zero. The numerically generated fields with the mean function $\mu(\mathbf{s}_i) = \beta = 20$, and the hyperparameter vector $\boldsymbol{\theta} = (\kappa, \alpha)^T$ being different values are shown in Fig. 7.2.

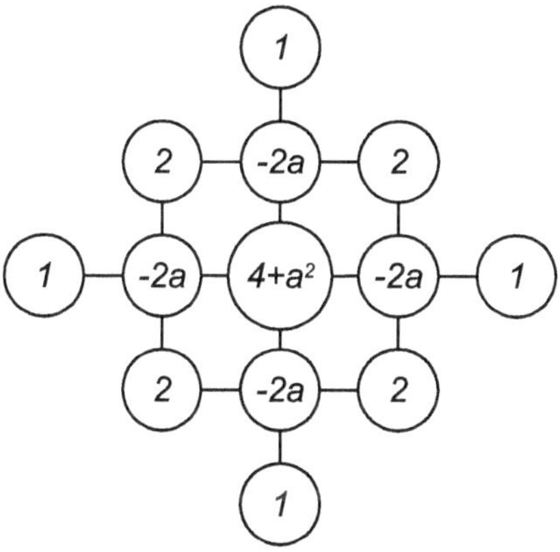

Fig. 7.1 Elements of the precision matrix $\mathbf{Q}$ related to a single location

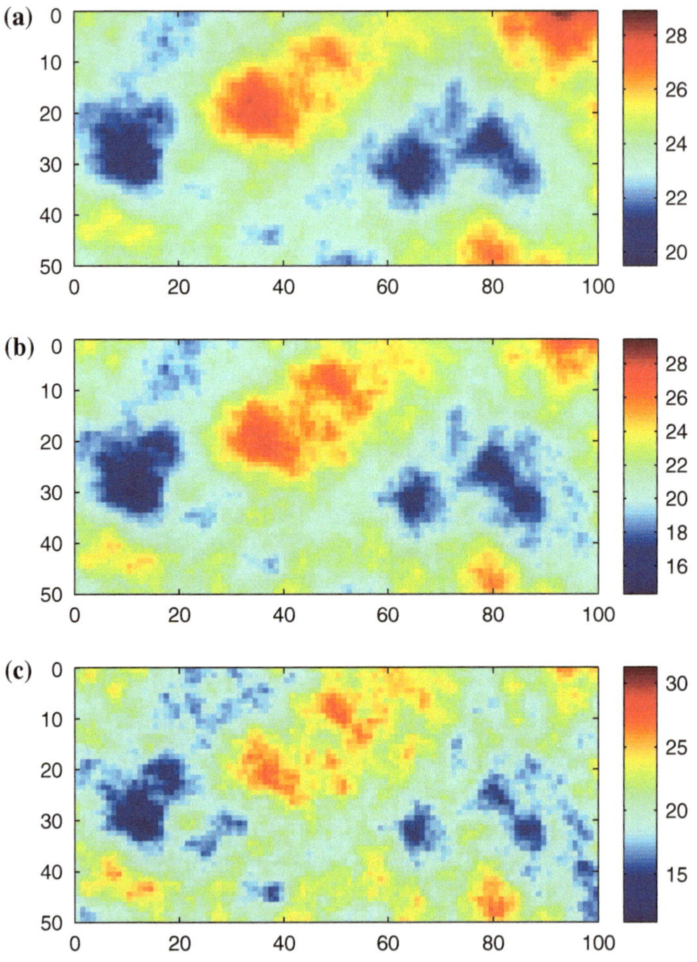

Fig. 7.2 Numerically generated spatial fields defined in (7.1) with $\mu(\mathbf{s}_i) = \beta = 20$, and $\mathbf{Q}_{\eta|\theta}$ constructed using (7.2) with hyperparameters being **a** $\boldsymbol{\theta} = (4, 0.0025)^T$, **b** $\boldsymbol{\theta} = (1, 0.01)^T$, and **c** $\boldsymbol{\theta} = (0.25, 0.04)^T$

7.2 Bayesian Predictive Inference

In this section, we propose a Bayesian inference approach to make predictive inferences of a spatial field $\mathbf{z}_* \in \mathbb{R}^{n_*}$.

First, we assign the vector of regression coefficients $\boldsymbol{\beta} \in \mathbb{R}^p$ with a Gaussian prior, namely $\boldsymbol{\beta} \sim \mathbb{N}\left(0, \mathbf{T}^{-1}\right)$, where the precision matrix $\mathbf{T} \in \mathbb{R}^{p \times p}$ is often chosen as a diagonal matrix with small diagonal elements when no prior information is available. Hence, the distribution of latent variables $\mathbf{z}$ given $\boldsymbol{\beta}$ and the hyperparameter vector $\boldsymbol{\theta}$ is Gaussian, i.e.,

$$\mathbf{z}|\beta, \theta \sim \mathbb{N}\left(\mathbf{F}\beta, \mathbf{Q}_{\eta|\theta}^{-1}\right),$$

where $\mathbf{F} = (f(\mathbf{s}_1), \ldots, f(\mathbf{s}_n))^T \in \mathbb{R}^{n \times p}$. For notational simplicity, we denote the full latent field of dimension $n + p$ by $\mathbf{x} = (\mathbf{z}^T, \beta^T)^T$. Then, for a given hyperparameter vector $\boldsymbol{\theta}$, the distribution $\pi(\mathbf{x}|\boldsymbol{\theta})$ is Gaussian obtained by

$$\begin{aligned}\pi(\mathbf{x}|\boldsymbol{\theta}) &= \pi(\mathbf{z}|\beta, \boldsymbol{\theta})\pi(\beta) \\ &\propto \exp\left(-\frac{1}{2}(\mathbf{z} - \mathbf{F}\beta)^T \mathbf{Q}_{\eta|\theta}(\mathbf{z} - \mathbf{F}\beta) - \frac{1}{2}\beta^T \mathbf{T}\beta\right) \\ &= \exp\left(-\frac{1}{2}\mathbf{x}^T \mathbf{Q}_{\mathbf{x}|\theta}\mathbf{x}\right),\end{aligned}$$

where the precision matrix $\mathbf{Q}_{\mathbf{x}|\theta} \in \mathbb{R}^{(n+p)\times(n+p)}$ is defined by

$$\mathbf{Q}_{\mathbf{x}|\theta} = \begin{bmatrix} \mathbf{Q}_{\eta|\theta} & -\mathbf{Q}_{\eta|\theta}\mathbf{F} \\ -\mathbf{F}^T\mathbf{Q}_{\eta|\theta} & \mathbf{F}^T\mathbf{Q}_{\eta|\theta}\mathbf{F} + \mathbf{T} \end{bmatrix}.$$

By the matrix inversion lemma, the covariance matrix $\boldsymbol{\Sigma}_{\mathbf{x}|\theta} \in \mathbb{R}^{(n+p)\times(n+p)}$ can be obtained by

$$\boldsymbol{\Sigma}_{\mathbf{x}|\theta} = \mathbf{Q}_{\mathbf{x}|\theta}^{-1} = \begin{bmatrix} \mathbf{Q}_{\eta|\theta}^{-1} + \mathbf{F}\mathbf{T}^{-1}\mathbf{F}^T & \mathbf{F}\mathbf{T}^{-1} \\ (\mathbf{F}\mathbf{T}^{-1})^T & \mathbf{T}^{-1} \end{bmatrix}.$$

At time $t \in \mathbb{Z}_{>0}$, we have a collection of observational data $\mathbf{y}_{1:t} \in \mathbb{R}^{Nt}$ obtained by the mobile sensing agents over time. Let $\mathbf{A}_{1:t} = (\mathbf{A}_1, \ldots, \mathbf{A}_t) \in \mathbb{R}^{(n+p)\times Nt}$, where $\mathbf{A}_\tau \in \mathbb{R}^{(n+p)\times N}$ is defined by

$$(\mathbf{A}_\tau)_{ij} = \begin{cases} 1, & \text{if } \mathbf{s}_i = \mathbf{q}_{\tau,j}, \\ 0, & \text{otherwise.} \end{cases}$$

Then the covariance matrix of $\mathbf{y}_{1:t}$ can be obtained by

$$\mathbf{R}_{1:t} = \mathbf{A}_{1:t}^T \boldsymbol{\Sigma}_{\mathbf{x}|\theta}\mathbf{A}_{1:t} + \mathbf{P}_{1:t},$$

where $\mathbf{P}_{1:t} = \sigma_w^2\mathbf{I} \in \mathbb{R}^{Nt\times Nt}$. By Gaussian process regression [53], the full conditional distribution of $\mathbf{x}$ is also Gaussian, i.e.,

$$\mathbf{x}|\theta, \mathbf{y}_{1:t} \sim \mathbb{N}(\mu_{\mathbf{x}|\theta,\mathbf{y}_{1:t}}, \boldsymbol{\Sigma}_{\mathbf{x}|\theta,\mathbf{y}_{1:t}}),$$

where

$$\begin{aligned}\boldsymbol{\Sigma}_{\mathbf{x}|\theta,\mathbf{y}_{1:t}} &= \boldsymbol{\Sigma}_{\mathbf{x}|\theta} - \boldsymbol{\Sigma}_{\mathbf{x}|\theta}\mathbf{A}_{1:t}\mathbf{R}_{1:t}^{-1}\mathbf{A}_{1:t}^T\boldsymbol{\Sigma}_{\mathbf{x}|\theta}, \\ \mu_{\mathbf{x}|\theta,\mathbf{y}_{1:t}} &= \boldsymbol{\Sigma}_{\mathbf{x}|\theta}\mathbf{A}_{1:t}\mathbf{R}_{1:t}^{-1}\mathbf{y}_{1:t}.\end{aligned} \tag{7.3}$$

The posterior distribution of the hyperparameter vector $\boldsymbol{\theta}$ can be obtained via

$$\pi(\boldsymbol{\theta}|\mathbf{y}_{1:t}) \propto \pi(\mathbf{y}_{1:t}|\boldsymbol{\theta})\pi(\boldsymbol{\theta}),$$

where the log likelihood function is defined by

$$\log \pi(\mathbf{y}_{1:t}|\boldsymbol{\theta}) = -\frac{1}{2}\mathbf{y}_{1:t}^T \mathbf{R}_{1:t}^{-1} \mathbf{y}_{1:t} - \frac{1}{2}\log\det \mathbf{R}_{1:t} - \frac{Nt}{2}\log 2\pi. \tag{7.4}$$

If a discrete prior on the hyperparameter vector $\boldsymbol{\theta}$ is chosen with a support $\boldsymbol{\Theta} = \{\boldsymbol{\theta}_1, \ldots, \boldsymbol{\theta}_L\}$, the posterior predictive distribution $\pi(\mathbf{x}|\mathbf{y}_{1:t})$ can be obtained by

$$\pi(\mathbf{x}|\mathbf{y}_{1:t}) = \sum_{\ell} \pi(\mathbf{x}|\boldsymbol{\theta}_\ell, \mathbf{y}_{1:t})\pi(\boldsymbol{\theta}_\ell|\mathbf{y}_{1:t}). \tag{7.5}$$

The predictive mean and variance then follow as

$$\begin{aligned} \mu_{x_i|\mathbf{y}_{1:t}} &= \sum_{\ell} \mu_{x_i|\boldsymbol{\theta}_\ell, \mathbf{y}_{1:t}} \pi(\boldsymbol{\theta}_\ell|\mathbf{y}_{1:t}), \\ \sigma^2_{x_i|\mathbf{y}_{1:t}} &= \sum_{\ell} \sigma^2_{x_i|\boldsymbol{\theta}_\ell, \mathbf{y}_{1:t}} \pi(\boldsymbol{\theta}_\ell|\mathbf{y}_{1:t}) + \sum_{\ell} (\mu_{x_i|\boldsymbol{\theta}_\ell, \mathbf{y}_{1:t}} - \mu_{x_i|\mathbf{y}_{1:t}})^2 \pi(\boldsymbol{\theta}_\ell|\mathbf{y}_{1:t}), \end{aligned} \tag{7.6}$$

where $\mu_{x_i|\boldsymbol{\theta}_\ell, \mathbf{y}_{1:t}}$ is the ith element in $\mu_{\mathbf{x}|\boldsymbol{\theta}_\ell, \mathbf{y}_{1:t}}$, and $\sigma^2_{x_i|\boldsymbol{\theta}_\ell, \mathbf{y}_{1:t}}$ is the ith diagonal element in $\boldsymbol{\Sigma}_{\mathbf{x}|\boldsymbol{\theta}_\ell, \mathbf{y}_{1:t}}$.

Remark 7.1 The discrete prior $\pi(\theta)$ greatly reduced the computational complexity in that it enables summation in (7.5) instead of numerical integration which has to be performed with a choice of continuous prior distribution. However, the computation of the full conditional distribution $\pi(\mathbf{x}|\boldsymbol{\theta}, \mathbf{y}_{1:t})$ in (7.3) and the likelihood $\pi(\mathbf{y}_{1:t}|\boldsymbol{\theta})$ (7.4) requires the inversion of the covariance matrix $\mathbf{R}_{1:t}$, whose size grows as the time t increases. Thus, the running time grows fast as new observations are collected and it will soon become intractable.

7.3 Sequential Bayesian Inference

In this section, we exploit the sparsity of the precision matrix, and propose a sequential Bayesian prediction algorithm which can be performed in constant time and fast enough even for a very large spatial field.

7.3.1 Update Full Conditional Distribution

First, we rewrite the full conditional distribution $\pi(\mathbf{x}|\theta, \mathbf{y}_{1:t})$ in terms of the sparse precision matrix $\mathbf{Q}_{\mathbf{x}|\theta}$ as follows

$$\mathbf{x}|\theta, \mathbf{y}_{1:t} \sim \mathbb{N}(\mu_{\mathbf{x}|\theta,\mathbf{y}_{1:t}}, \mathbf{Q}^{-1}_{\mathbf{x}|\theta,\mathbf{y}_{1:t}}),$$

where

$$\begin{aligned} \mathbf{Q}_{\mathbf{x}|\theta,\mathbf{y}_{1:t}} &= \mathbf{Q}_{\mathbf{x}|\theta} + \mathbf{A}_{1:t}\mathbf{P}^{-1}_{1:t}\mathbf{A}^T_{1:t} \\ \mu_{\mathbf{x}|\theta,\mathbf{y}_{1:t}} &= \mathbf{Q}^{-1}_{\mathbf{x}|\theta,\mathbf{y}_{1:t}}\mathbf{A}_{1:t}\mathbf{P}^{-1}_{1:t}\mathbf{y}_{1:t}. \end{aligned} \tag{7.7}$$

From here on, we will use $\mathbf{Q}_{t|\theta} = \mathbf{Q}_{\mathbf{x}|\theta,\mathbf{y}_{1:t}}$ and $\mu_{t|\theta} = \mu_{\mathbf{x}|\theta,\mathbf{y}_{1:t}}$, for notational simplicity. Notice that (7.7) can be represented by the following recursion

$$\begin{aligned} \mathbf{Q}_{t|\theta} &= \mathbf{Q}_{t-1|\theta} + \frac{1}{\sigma_w^2}\sum_{i=1}^{N}\mathbf{u}_{t,i}\mathbf{u}^T_{t,i}, \\ \mathbf{b}_t &= \mathbf{b}_{t-1} + \frac{1}{\sigma_w^2}\sum_{i=1}^{N}\mathbf{u}_{t,i}y_{t,i}, \end{aligned} \tag{7.8}$$

where $\mathbf{b}_t = \mathbf{Q}_{t|\theta}\mu_{t|\theta}$ with initial conditions

$$\mathbf{Q}_{0|\theta} = \mathbf{Q}_{\mathbf{x}|\theta,\mathbf{y}_{1:0}} = \mathbf{Q}_{\mathbf{x}|\theta}, \text{ and } \mathbf{b}_0 = \mathbf{0}.$$

In (7.8), we have defined $\mathbf{u}_{t,i} \in \mathbb{R}^{n+p}$ as

$$(\mathbf{u}_{t,i})_j = \begin{cases} 1, & \text{if } \mathbf{s}_j = \mathbf{q}_{t,i}, \\ 0, & \text{otherwise.} \end{cases}$$

Lemma 7.1 *For a given $\theta \in \boldsymbol{\theta}$, the full conditional mean and variance, i.e., $\mu_{t|\theta}$ and $\mathbf{Q}_{t|\theta}$, can be updated in short constant time given $\mathbf{Q}_{t-1|\theta}$ and $\mathbf{b}_{t-1}$.*

Proof The update of $\mathbf{Q}_{t|\theta}$ and $\mathbf{b}_t$ can be obviously computed in constant time. Hence $\mu_{t|\theta}$ can be obtained by solving a linear equation $\mathbf{Q}_{t|\theta}\mu_{t|\theta} = \mathbf{b}_t$. Due to the sparse structure of $\mathbf{Q}_{t|\theta}$, this operation can be done in a very short time. Moreover, notice that $\mathbf{Q}_{t|\theta}$ and $\mathbf{Q}_{t-1|\theta}$ have the same sparsity structure and hence the computational complexity remains fixed. □

From Lemma 7.1, we can compute $\mu_{x_i|\theta,\mathbf{y}_{1:t}}$ in (7.6) in constant time. In order to find $\sigma^2_{x_i|\theta,\mathbf{y}_{1:t}}$ in (7.6), we need to compute $\mathbf{\Sigma}_{x|\theta,\mathbf{y}_{1:t}}$ which requires the inversion of $\mathbf{Q}_{t|\theta}$. The inversion of a big matrix (even a sparse matrix) is undesirable. However, notice that only the diagonal elements in $\mathbf{Q}^{-1}_{t|\theta}$ are needed. Following the Sherman-Morrison formula (see Appendix A.2.2) and using (7.8), $\sigma^2_{x_i|\theta,\mathbf{y}_{1:t}}$ can be obtained exactly via

$$
\begin{aligned}
\operatorname{diag}(\mathbf{Q}_{t|\theta}^{-1}) &= \operatorname{diag}\left(\left[\mathbf{Q}_{t-1|\theta} + \sum_{i=1}^{N} \mathbf{u}_{t,i}\mathbf{u}_{t,i}^T\right]^{-1}\right) \\
&= \operatorname{diag}(\mathbf{Q}_{t-1|\theta}^{-1}) - \sum_{i=1}^{N} \frac{\mathbf{h}_{t,i|\theta} \circ \mathbf{h}_{t,i|\theta}}{\sigma_w^2 + \mathbf{u}_{t,i}^T \mathbf{h}_{t,i|\theta}}, \qquad (7.9) \\
\mathbf{h}_{t,i|\theta} &= \mathbf{B}_{t,i|\theta}^{-1} \mathbf{u}_{t,i}, \\
\mathbf{B}_{t,i|\theta} &= \mathbf{Q}_{t-1|\theta} + \frac{1}{\sigma_w^2} \sum_{j=1}^{i} \mathbf{u}_{t,j}\mathbf{u}_{t,j}^T,
\end{aligned}
$$

where $\circ$ denotes the element-wise produce. By this way, the computation can be done efficiently in constant time.

7.3.2 Update Likelihood

Next, we derive the update rule for the log likelihood function. We have the following proposition.

Proposition 7.1 *The log likelihood function* $\log \pi(\mathbf{y}_{1:t}|\boldsymbol{\theta})$ *in* (7.4) *can be obtained by*

$$
\log \pi(\mathbf{y}_{1:t}|\boldsymbol{\theta}) = c_t + g_{t,\theta} + \frac{1}{2}\mathbf{b}_t^T \mu_{t|\theta} - \frac{Nt}{2}\log(2\pi\sigma_w^2) \qquad (7.10)
$$

where

$$
\begin{aligned}
c_t &= c_{t-1} - \frac{1}{2\sigma_w^2}\sum_{i=1}^{N} y_{t,i}^2, \quad c_0 = 0, \\
g_{t|\theta} &= g_{t-1|\theta} - \frac{1}{2}\sum_{i=1}^{N} \log\left(1 + \frac{1}{\sigma_w^2}\mathbf{u}_{t,i}^T \mathbf{h}_{t,i|\theta}\right), \quad g_{0|\theta} = 0,
\end{aligned}
$$

with $\mathbf{h}_{t,i|\theta}$ *defined in* (7.9).

Proof The inverse of the covariance matrix $\mathbf{R}_{1:t}$ can be obtained by

$$
\begin{aligned}
\mathbf{R}_{1:t}^{-1} &= (\mathbf{A}_{1:t}^T \mathbf{Q}_{0|\theta}^{-1} \mathbf{A}_{1:t} + \mathbf{P}_{1:t})^{-1} \\
&= \mathbf{P}_{1:t}^{-1} - \mathbf{P}_{1:t}^{-1}\mathbf{A}_{1:t}^T(\mathbf{Q}_{0|\theta} + \mathbf{A}_{1:t}\mathbf{P}_{1:t}^{-1}\mathbf{A}_{1:t}^T)^{-1}\mathbf{A}_{1:t}\mathbf{P}_{1:t}^{-1} \\
&= \mathbf{P}_{1:t}^{-1} - \mathbf{P}_{1:t}^{-1}\mathbf{A}_{1:t}^T\mathbf{Q}_{t|\theta}^{-1}\mathbf{A}_{1:t}\mathbf{P}_{1:t}^{-1}.
\end{aligned}
$$

Similarly, the log determinant of the covariance matrix $\boldsymbol{\Sigma}_{1:t}$ can be obtained by

$$\begin{aligned}
\log\det \mathbf{R}_{1:t} &= \log\det(\mathbf{A}_{1:t}^T \mathbf{Q}_{0|\theta}^{-1} \mathbf{A}_{1:t} + \mathbf{P}_{1:t}) \\
&= \log\det(\mathbf{I} + \frac{1}{\sigma_w^2} \mathbf{A}_{1:t}^T \mathbf{Q}_{0|\theta}^{-1} \mathbf{A}_{1:t}) + Nt \log \sigma_w^2 \\
&= \log\det(\mathbf{Q}_{0|\theta} + \frac{1}{\sigma_w^2} \sum_{\tau=1}^{t} \sum_{i=1}^{N} \mathbf{u}_{\tau,i} \mathbf{u}_{\tau,i}^T) - \log\det(\mathbf{Q}_{0|\theta}) + Nt \log \sigma_w^2 \\
&= \sum_{\tau=1}^{t} \log(1 + \mathbf{u}_\tau^T \mathbf{Q}_{\tau-1|\theta}^{-1} \mathbf{u}_\tau) + Nt \log \sigma_w^2 .
\end{aligned}$$

Hence, we have

$$\begin{aligned}
&\log \pi(\mathbf{y}_{1:t}|\boldsymbol{\theta}) \\
&= -\tfrac{1}{2} \mathbf{y}_{1:t}^T \mathbf{R}_{1:t}^{-1} \mathbf{y}_{1:t} - \tfrac{1}{2} \log\det \mathbf{R}_{1:t} - \tfrac{Nt}{2} \log 2\pi \\
&= -\tfrac{1}{2} \mathbf{y}_{1:t}^T \mathbf{P}_{1:t}^{-1} \mathbf{y}_{1:t} + \tfrac{1}{2} \mathbf{b}_t^T \mu_{t|\theta} - \tfrac{1}{2} \sum_{\tau=1}^{t} \sum_{i=1}^{N} \log(1 + \mathbf{u}_{\tau,i}^T \mathbf{B}_{\tau,i|\theta}^{-1} \mathbf{u}_{\tau,i}) - \tfrac{Nt}{2} \log(2\pi\sigma_w^2).
\end{aligned}$$

□

Lemma 7.2 *For a given $\boldsymbol{\theta} \in \boldsymbol{\theta}$, the log likelihood function, i.e.,* $\log \pi(\mathbf{y}_{1:t}|\boldsymbol{\theta})$ *can be computed in short constant time.*

Proof The result follows directly from Proposition 7.1. □

7.3.3 Update Predictive Distribution

Combining the results in Lemmas 7.1, 7.2, and (7.5), (7.6), we summarize our results in the following theorem.

Theorem 7.1 *The predictive distribution in* (7.5) *(or the predictive mean and variance in* (7.6)*) can be obtained in constant time as time t increases.*

We summarize the proposed sequential Bayesian prediction algorithm in Table 7.1.

Table 7.1 Sequential Bayesian predictive inference

Input:
(1) prior distribution of $\boldsymbol{\theta} \in \boldsymbol{\Theta}$, $i.e.$, $\pi(\theta)$
Output:
(1) predictive mean $\left\{\mu_{x_i \mid \mathbf{y}_{1:t}}\right\}_{i=1}^{n_*}$
(2) predictive variance $\left\{\sigma_{x_i \mid \mathbf{y}_{1:t}}^2\right\}_{i=1}^{n_*}$

Initialization:
1: initialize $\mathbf{b} = \mathbf{0}$, $c = 0$
2: **for** $\boldsymbol{\theta} \in \boldsymbol{\Theta}$ **do**
3:     initialize $\mathbf{Q}_{\boldsymbol{\theta}}$, $g_{\boldsymbol{\theta}} = 0$
4:     compute $\mathrm{diag}(\mathbf{Q}_{\boldsymbol{\theta}}^{-1})$
5: **end for**
At time $t \in \mathbb{Z}_{>0}$, do:
1: **for** $1 \leq i \leq N$ **do**
2:     obtain new observations $y_{t,i}$ collected at current locations $\mathbf{q}_{t,i}$
3:     find the index k corresponding to $\mathbf{q}_{t,i}$, and set $\mathbf{u} = \mathbf{e}_k$
4:     update $\mathbf{b} = \mathbf{b} + \frac{y_{t,i}}{\sigma_w^2}\mathbf{u}$
5:     update $c = c - \frac{1}{2\sigma_w^2}y_{t,i}^2$
6:     **for** $\boldsymbol{\theta} \in \boldsymbol{\Theta}$ **do**
7:         compute $\mathbf{h}_{\boldsymbol{\theta}} = \mathbf{Q}_{\boldsymbol{\theta}}^{-1}\mathbf{u}$
8:         update $\mathrm{diag}(\mathbf{Q}_{\boldsymbol{\theta}}^{-1}) = \mathrm{diag}(\mathbf{Q}_{\boldsymbol{\theta}}^{-1}) - \frac{\mathbf{h}_{\boldsymbol{\theta}} \circ \mathbf{h}_{\boldsymbol{\theta}}}{\sigma_w^2 + \mathbf{u}^T \mathbf{h}_{\boldsymbol{\theta}}}$
9:         update $\mathbf{Q}_{\boldsymbol{\theta}}$ via $\mathbf{Q}_{\boldsymbol{\theta}} = \mathbf{Q}_{\boldsymbol{\theta}} + \frac{1}{\sigma_w^2}\mathbf{u}\mathbf{u}^T$
10:       update $g_{\boldsymbol{\theta}} = g_{\boldsymbol{\theta}} - \frac{1}{2}\log(1 + \frac{1}{\sigma_w^2}\mathbf{u}^T\mathbf{h})$
11:    **end for**
12: **end for**
13: **for** $\boldsymbol{\theta} \in \boldsymbol{\Theta}$ **do**
14:     compute $\boldsymbol{\mu}_{\boldsymbol{\theta}} = \mathbf{Q}_{\boldsymbol{\theta}}^{-1}\mathbf{b}$
15:     compute the likelihood via
$$\log \pi(\boldsymbol{\theta}\mid\mathbf{y}_{1:t}) = c + g_{\boldsymbol{\theta}} + \tfrac{1}{2}\mathbf{b}^T\boldsymbol{\mu}_{\boldsymbol{\theta}}$$
16: **end for**
17: compute the posterior distribution via
$$\pi(\boldsymbol{\theta}\mid\mathbf{y}_{1:t}) \propto \pi(\mathbf{y}_{1:t}\mid\boldsymbol{\theta})\pi(\boldsymbol{\theta})$$
18: compute the predictive mean via
$$\mu_{x_i\mid\mathbf{y}_{1:t}} = \textstyle\sum_{\ell}(\boldsymbol{\mu}_{\boldsymbol{\theta}_\ell})_i \pi(\boldsymbol{\theta}_\ell\mid\mathbf{y}_{1:t})$$
19: compute the predictive variance via
$$\sigma_{x_i\mid\mathbf{y}_{1:t}}^2 = \textstyle\sum_{\ell}\left((\mathrm{diag}(\mathbf{Q}_{\boldsymbol{\theta}_\ell}))_i + ((\boldsymbol{\mu}_{\boldsymbol{\theta}_\ell})_i - \mu_{x_i\mid\mathbf{y}_{1:t}})^2\right)\pi(\boldsymbol{\theta}_\ell\mid\mathbf{y}_{1:t})$$

7.4 Adaptive Sampling

In the previous section, we have designed a sequential Bayesian prediction algorithm for estimating the scalar field at time t. In this section, we propose an adaptive sampling strategy for finding most informative sampling locations at time $t+1$ for mobile sensing agents in order to improve the quality of prediction and reduce the uncertainty in hyper parameters simultaneously.

In our previous work [118], we have proposed to use the conditional entropy $H(\mathbf{z}_*|\boldsymbol{\theta} = \hat{\boldsymbol{\theta}}_t, \mathbf{y}_{1:t+1})$ as an optimality criterion, where

$$\hat{\boldsymbol{\theta}}_t = \arg\max_{\boldsymbol{\theta}} \pi(\boldsymbol{\theta}|\mathbf{y}_{1:t}),$$

is the maximum *a posterior* (MAP) estimate based on the cumulative observations up to current time t. Although this approach greatly simplifies the computation, it does not count for the uncertainty in estimating the hyperparameter vector $\boldsymbol{\theta}$.

In this chapter, we propose to use the conditional entropy $H(\mathbf{z}_*, \boldsymbol{\theta}|\mathbf{y}_{t+1}, \mathbf{y}_{1:t})$ which represents the uncertainty remained in both random vectors $\mathbf{z}_*$ and $\boldsymbol{\theta}$ by knowing future measurements in the random vector $\mathbf{y}_{t+1}$. Notice that the measurements $\mathbf{y}_{1:t}$ have been observed and treated as constants. It can be obtained by

$$\begin{aligned} H(\mathbf{z}_*, \boldsymbol{\theta}|\mathbf{y}_{t+1}, \mathbf{y}_{1:t}) &= H(\mathbf{z}_*|\boldsymbol{\theta}, \mathbf{y}_{t+1}, \mathbf{y}_{1:t}) + H(\boldsymbol{\theta}|\mathbf{y}_{t+1}, \mathbf{y}_{1:t}) \\ &= H(\mathbf{z}_*|\boldsymbol{\theta}, \mathbf{y}_{t+1}, \mathbf{y}_{1:t}) + H(\mathbf{y}_{t+1}|\boldsymbol{\theta}, \mathbf{y}_{1:t}) \\ &\quad + H(\boldsymbol{\theta}|\mathbf{y}_{1:t}) - H(\mathbf{y}_{t+1}|\mathbf{y}_{1:t}). \end{aligned}$$

Notice that we have the following Gaussian distributions (the means will not be exploited and hence not shown here):

$$\begin{aligned} \mathbf{z}_*|\boldsymbol{\theta}, \mathbf{y}_{t+1}, \mathbf{y}_{1:t} &\sim \mathbb{N}(\cdot, \boldsymbol{\Sigma}_{\mathbf{x}_*|\boldsymbol{\theta}, \mathbf{y}_{1:t+1}}), \\ \mathbf{y}_{t+1}|\boldsymbol{\theta}, \mathbf{y}_{1:t} &\sim \mathbb{N}(\cdot, \boldsymbol{\Sigma}_{\mathbf{y}_{t+1}|\boldsymbol{\theta}, \mathbf{y}_{1:t}} + \sigma_w^2 \mathbf{I}), \\ \mathbf{y}_{t+1}|\mathbf{y}_{1:t} &\overset{\text{approx}}{\sim} \mathbb{N}(\cdot, \boldsymbol{\Sigma}_{\mathbf{y}_{t+1}|\mathbf{y}_{1:t}} + \sigma_w^2 \mathbf{I}), \end{aligned}$$

in which the last one is approximated using (7.6). Notice that the approximation is used here to avoid numerical integration over the random vector $\mathbf{y}_{t+1}$ which needs to be done using Monte Carlo methods. Moreover, the entropy $H(\boldsymbol{\theta}|\mathbf{y}_{1:t}) = c$ is a constant since $\mathbf{y}_{1:t}$ is known. Since the entropy for a multivariate Gaussian distribution has a closed-from expression [72], we have

$$
\begin{aligned}
H(\mathbf{z}_*, \theta | \mathbf{y}_{t+1}, \mathbf{y}_{1:t}) = & \sum_{\ell} \frac{1}{2} \log \left((2\pi e)^{n_*} \det(\boldsymbol{\Sigma}_{\mathbf{x}_* | \theta_\ell, \mathbf{y}_{1:t+1}}) \right) \pi(\theta_\ell | \mathbf{y}_{1:t}) \\
& + \sum_{\ell} \frac{1}{2} \log \left((2\pi e)^{N} \det(\boldsymbol{\Sigma}_{\mathbf{q}_{t+1} | \theta_\ell, \mathbf{y}_{1:t}}) \right) \pi(\theta_\ell | y_{1:t}) \\
& - \frac{1}{2} \log \left((2\pi e)^{N} \det(\boldsymbol{\Sigma}_{\mathbf{q}_{t+1} | \mathbf{y}_{1:t}}) \right) + c.
\end{aligned}
$$

It can also be shown that

$$
\begin{aligned}
\log\det(\boldsymbol{\Sigma}_{\mathbf{x}_* | \theta_\ell, \mathbf{y}_{1:t+1}}) &= \log\det(\mathbf{Q}^{-1}_{t+1|\theta_\ell})_{(\mathcal{S}_*)} \\
&= \log\det(\mathbf{Q}_{t+1|\theta_\ell})_{(-\mathcal{S}_*)} - \log\det(\mathbf{Q}_{t+1|\theta_\ell}),
\end{aligned}
$$

where $\mathbf{A}_{(\mathcal{S}_*)}$ denotes the submatrix of $\mathbf{A}$ formed by the first 1 to n_* rows and columns (recall that $\mathcal{S}_* = \{\mathbf{s}_1, \ldots, \mathbf{s}_{n_*}\}$). Notice that the term $\log\det(\mathbf{Q}_{t+1|\theta_\ell})_{(-\mathcal{S}_*)}$ is a constant since agents only sample at $\mathcal{S}_*$. Hence, the optimal sampling locations at time $t+1$ can be determined by solving the following optimization problem

$$
\begin{aligned}
\mathbf{q}^*_{t+1} &= \arg \min_{\{\mathbf{q}_{t+1,i} \in \mathcal{R}_{t,i}\}} H(\mathbf{z}_*, \theta | \mathbf{y}_{t+1}, \mathbf{y}_{1:t}) \\
&= \arg \min_{\{\mathbf{q}_{t+1,i} \in \mathcal{R}_{t,i}\}} \sum_{\ell} - \log\det(\mathbf{Q}_{t+1|\theta_\ell}) \pi(\theta_\ell | \mathbf{y}_{1:t}) \\
& \quad + \sum_{\ell} \log\det(\boldsymbol{\Sigma}_{\mathbf{y}_{t+1} | \theta_\ell, \mathbf{y}_{1:t}}) \pi(\theta_\ell | \mathbf{y}_{1:t}) - \log\det(\boldsymbol{\Sigma}_{\mathbf{y}_{t+1} | \mathbf{y}_{1:t}}),
\end{aligned}
$$

where $\mathcal{R}_{t,i} = \{\mathbf{s} | \ \|\mathbf{s} - \mathbf{q}_{t,i}\| \le r, \mathbf{s} \in \mathcal{S}_*\}$ (in which $r \in \mathbb{R}_{>0}$ is the maximum distance an agent can move between time instances) is the reachable set at time t. This combinatorial optimization problem can be solved using a greedy algorithm, i.e., finding the sub-optimal sampling locations for agents in sequence.

7.5 Simulation

In this section, we demonstrate the effectiveness of the proposed sequential Bayesian inference algorithm and the adaptive sampling strategy through a numerical experiment.

Consider a spatial field introduced in Example 7.1. The mean function is a constant $\beta = 20$. We choose the precision matrix $\mathbf{Q}_{\mathbf{x}|\theta}$ with hyperparameters $\alpha = 0.01$ equivalent to a bandwidth $\ell = 1/\sqrt{\alpha} = 10$, and $\kappa = 1$ equivalent to a vertical scale $\sigma_f^2 = 1/4\pi\alpha\kappa \approx 8$. The numerically generated field is shown in Fig. 7.2b. The precision matrix $T \in \mathbb{R}$ of β is chosen to be 10^{-4}. The measurement noise level $\sigma_w = 0.2$ is assumed to be known. A discrete uniform distribution is selected with a support shown in Fig. 7.3. $N = 5$ mobile sensing agents take measurements at time

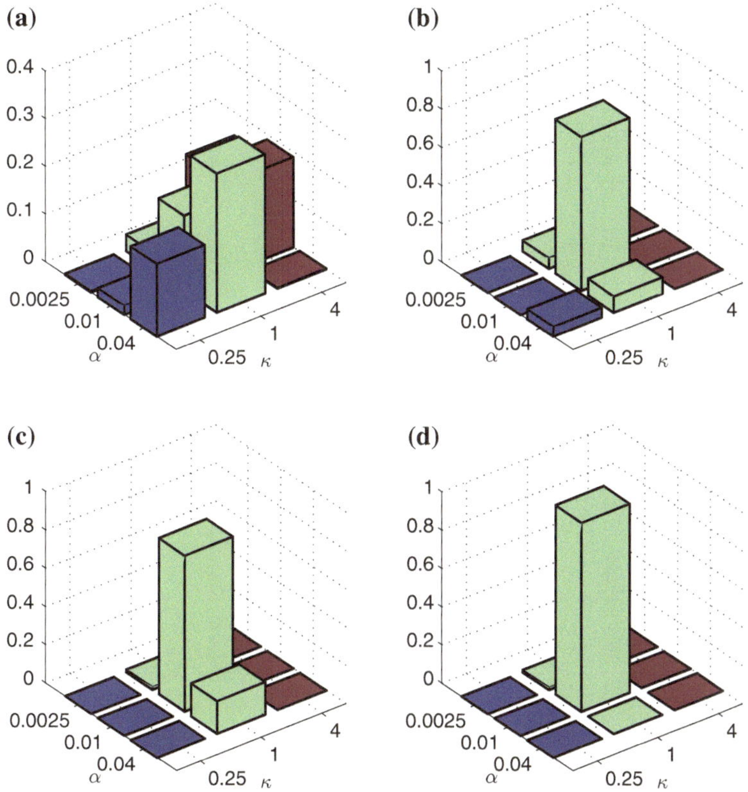

Fig. 7.3 Posterior distributions of θ, i.e., $\pi(\boldsymbol{\theta}|\mathbf{y}_{1:t})$, at **a** $t = 1$, **b** $t = 5$, **c** $t = 10$, and **d** $t = 20$

$t \in \mathbb{Z}_{>0}$, starting from locations shown in Fig. 7.4b (in white dots). The maximum distance each agent can travel between time instances is chosen to be $r = 5$.

Figure 7.4 shows the predicted fields and the prediction error variances at times $t = 1, 5, 10, 20$. The trajectories of agents are shown in white circles with the current locations shown in white dots. It can be seen that agents try to cover the field of interest as time evolves. The predicted field (the predictive mean) gets closer to the true field (see Fig. 7.2b) and the prediction error variances become smaller as more observations are collected. Figure 7.3 shows the posterior distribution of the hyperparameters in $\boldsymbol{\theta}$. Clearly, as more measurements are obtained, this posterior distribution becomes peaked at the true value (1,0.01). Figure 7.5a shows the predicted distribution of the estimated mean β as time evolves. In Fig. 7.5b, we can see that the RMS error computed via

$$\mathrm{rms}(t) = \sqrt{\frac{1}{n_*}\sum_{i=1}^{n_*}(\mu_{z_i|\mathbf{y}_{1:t}} - z_i)^2},$$

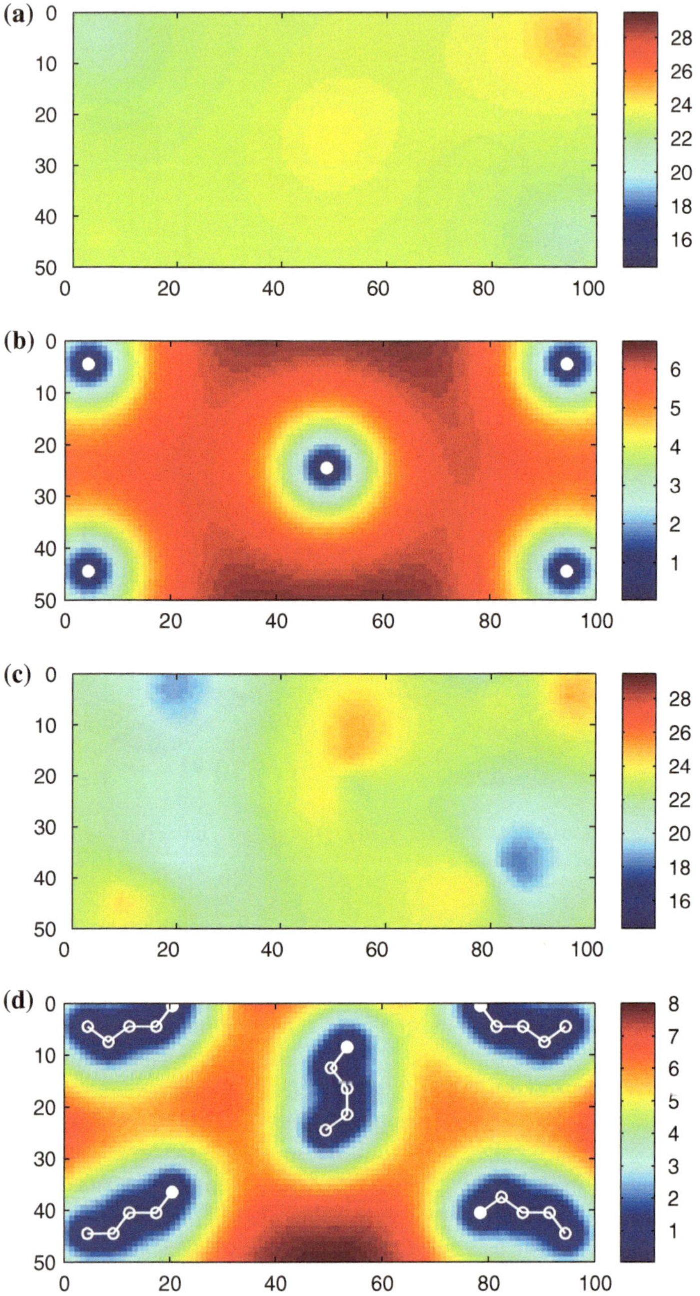

Fig. 7.4 Predicted fields at **a** $t = 1$, **c** $t = 5$, **e** $t = 10$, and **g** $t = 20$. Prediction error variances at **b** $t = 1$, **d** $t = 5$, **f** $t = 10$, and **h** $t = 20$

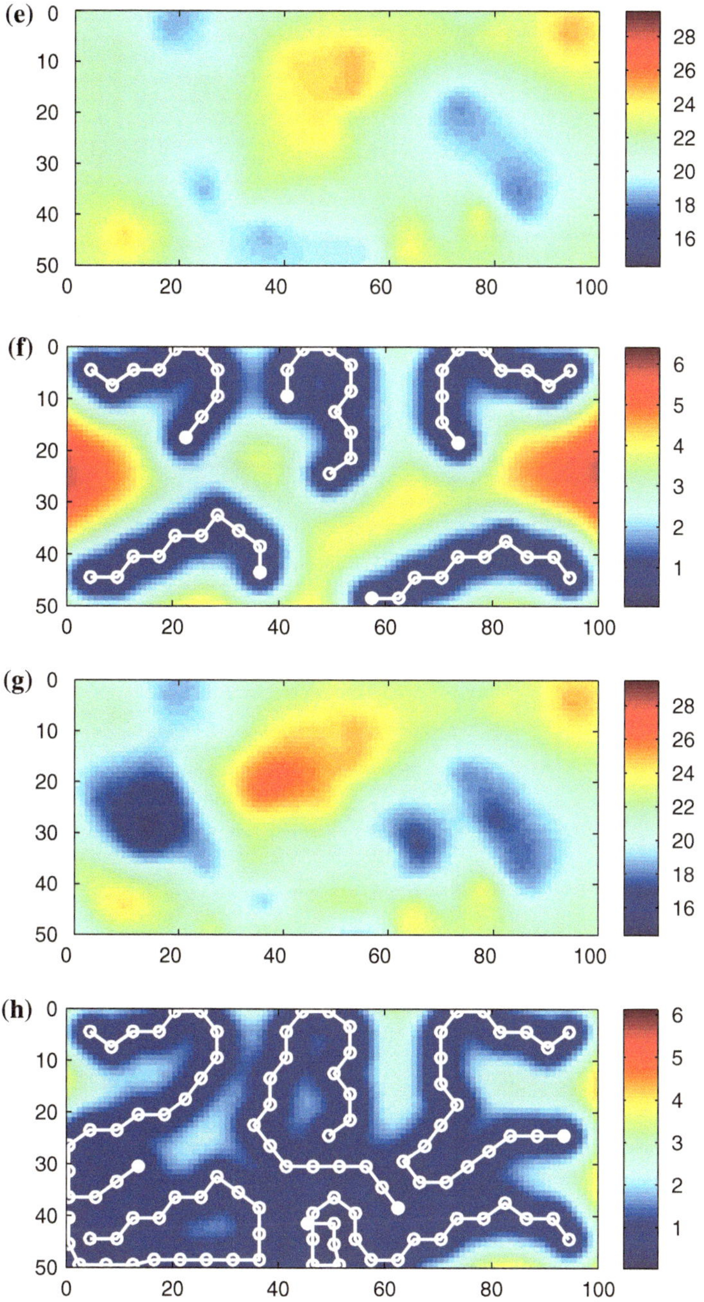

Fig. 7.4 (continued)

Fig. 7.5 **a** Estimated β, and **b** root mean square error

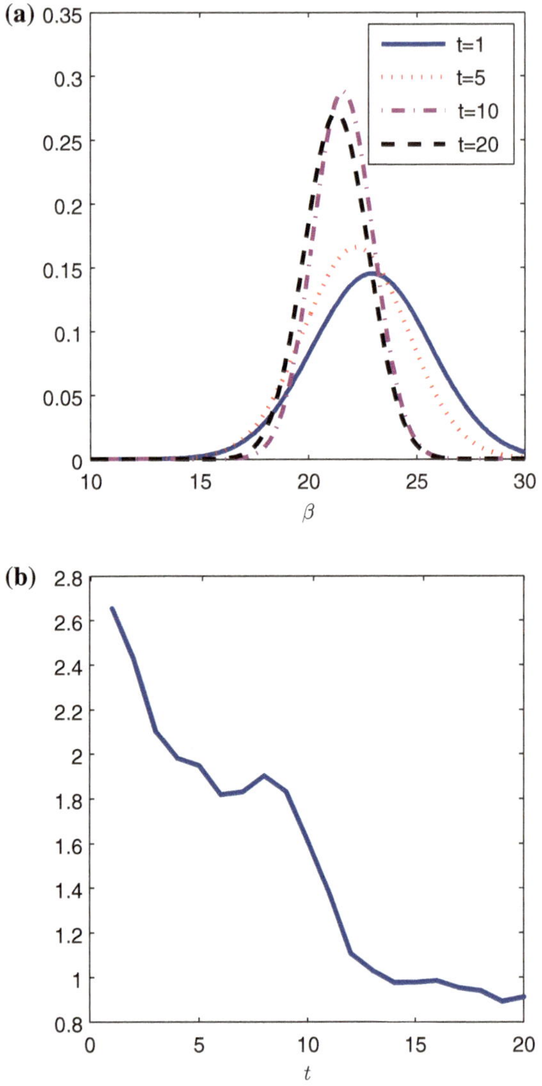

decreases as time increases, which shows the effectiveness of the proposed scheme.

The most important contribution is that the computation time at each time step does not grow as the number of measurements increases.

Appendix A
Mathematical Background

A.1 Gaussian Identities

The multivariate Gaussian distribution of a random vector $\mathbf{x} \in \mathbb{R}^n$ (i.e., $\mathbf{x} \sim \mathbb{N}(\boldsymbol{\mu}, \boldsymbol{\Sigma})$) has a joint probability density function (pdf) given by

$$p(\mathbf{x}; \boldsymbol{\mu}, \boldsymbol{\Sigma}) = \frac{1}{(2\pi)^{-n/2}|\boldsymbol{\Sigma}|^{-1/2}} \exp\left(-\frac{1}{2}(\mathbf{x} - \boldsymbol{\mu})^T \boldsymbol{\Sigma}^{-1}(\mathbf{x} - \boldsymbol{\mu})\right),$$

where $\boldsymbol{\mu} \in \mathbb{R}^n$ is the mean vector, and $\boldsymbol{\Sigma} \in \mathbb{R}^{n \times n}$ is the covariance matrix.

Now, suppose $\mathbf{x}$ consists of two disjoint subsets $\mathbf{x}_a$ and $\mathbf{x}_b$, i.e.,

$$\mathbf{x} = \begin{bmatrix} \mathbf{x}_a \\ \mathbf{x}_b \end{bmatrix}.$$

The corresponding mean vector μ and covariance matrix Σ can be written as

$$\boldsymbol{\mu} = \begin{bmatrix} \boldsymbol{\mu}_a \\ \boldsymbol{\mu}_b \end{bmatrix}, \quad \Sigma = \begin{bmatrix} \boldsymbol{\Sigma}_{aa} & \boldsymbol{\Sigma}_{ab} \\ \boldsymbol{\Sigma}_{ba} & \boldsymbol{\Sigma}_{bb} \end{bmatrix},$$

where $\boldsymbol{\Sigma}_{ab} = \boldsymbol{\Sigma}_{ba}^T$ due to the symmetry of $\boldsymbol{\Sigma}$. Then, the marginal distribution of $\mathbf{x}_a$ is given by

$$\mathbf{x}_a \sim \mathbb{N}(\boldsymbol{\mu}_a, \boldsymbol{\Sigma}_{aa}),$$

and the conditional distribution of $\mathbf{x}_a$ given $\mathbf{x}_b$ is given by

$$\mathbf{x}_a|\mathbf{x}_b \sim \mathbb{N}(\boldsymbol{\mu}_{a|b}, \boldsymbol{\Sigma}_{a|b}),$$

where

$$\boldsymbol{\mu}_{a|b} = \boldsymbol{\mu}_a + \boldsymbol{\Sigma}_{ab}\boldsymbol{\Sigma}_{bb}^{-1}(\mathbf{x}_b - \boldsymbol{\mu}_b)$$
$$\boldsymbol{\Sigma}_{a|b} = \boldsymbol{\Sigma}_{aa} - \boldsymbol{\Sigma}_{ab}\boldsymbol{\Sigma}_{bb}^{-1}\boldsymbol{\Sigma}_{ba}.$$

Y. Xu et al., *Bayesian Prediction and Adaptive Sampling Algorithms for Mobile Sensor Networks*, SpringerBriefs in Control, Automation and Robotics, DOI 10.1007/978-3-319-21921-9

A.2 Matrix Inversion Lemma

Matrices can be inverted blockwise by using the following analytic inversion formula:

$$\begin{bmatrix} \mathbf{A} & \mathbf{B} \\ \mathbf{B}^T & \mathbf{C} \end{bmatrix}^{-1} = \begin{bmatrix} \mathbf{A}^{-1} + \mathbf{A}^{-1}\mathbf{B}(\mathbf{C} - \mathbf{B}^T\mathbf{A}^{-1}\mathbf{B})^{-1}\mathbf{B}^T\mathbf{A}^{-1} & -\mathbf{A}^{-1}\mathbf{B}(\mathbf{C} - \mathbf{B}^T\mathbf{A}^{-1}\mathbf{B})^{-1} \\ -(\mathbf{C} - \mathbf{B}^T\mathbf{A}^{-1}\mathbf{B})^{-1}\mathbf{B}^T\mathbf{A}^{-1} & (\mathbf{C} - \mathbf{B}^T\mathbf{A}^{-1}\mathbf{B})^{-1} \end{bmatrix},$$

where $\mathbf{A}$, $\mathbf{B}$, and $\mathbf{C}$ are matrix subblocks of arbitrary size. Matrices $\mathbf{A}$ and $\mathbf{C} - \mathbf{B}^T\mathbf{A}^{-1}\mathbf{B}$ must be nonsingular.

A.2.1 Woodbury Identity

The Woodbury matrix identity is

$$(\mathbf{A} + \mathbf{UCV})^{-1} = \mathbf{A}^{-1} - \mathbf{A}^{-1}\mathbf{U}\left(\mathbf{C}^{-1} + \mathbf{VA}^{-1}\mathbf{U}\right)^{-1}\mathbf{VA}^{-1},$$

where $\mathbf{A}$, $\mathbf{U}$, $\mathbf{C}$, and $\mathbf{V}$ denote matrices with appropriate size.

A.2.2 Sherman–Morrison Formula

Suppose $\mathbf{A} \in \mathbb{R}^{n\times n}$ is invertible and $\mathbf{u} \in \mathbb{R}^n$ and $\mathbf{v} \in \mathbb{R}^n$ are vectors. Assume that $1 + \mathbf{v}^T\mathbf{A}^{-1}\mathbf{u} \neq 0$, the Sherman–Morrison formula states that

$$(\mathbf{A} + \mathbf{uv}^T)^{-1} = \mathbf{A}^{-1} - \frac{\mathbf{A}^{-1}\mathbf{uv}^T\mathbf{A}^{-1}}{1 + \mathbf{v}^T\mathbf{A}^{-1}\mathbf{u}}.$$

A.3 Generating Gaussian Processes

In order to implement algorithms in simulation studies, we need to generate multivariate Gaussian samples from $\mathbb{N}(\boldsymbol{\mu}, \boldsymbol{\Sigma})$ with arbitrary mean $\boldsymbol{\mu}$ and covariance matrix $\boldsymbol{\Sigma}$. In what follows, we introduce two approaches.

A.3.1 Cholesky Decomposition

Given an arbitrary mean $\boldsymbol{\mu}$ and a positive definite covariance matrix $\boldsymbol{\Sigma}$, the algorithm generates multivariate Gaussian samples is shown in Table A.1.

A.3.2 Circulant Embedding

Consider a 1D zero-mean stationary Gaussian process $z(x)$ with a covariance function $C(x, x')$. The covariance matrix $\boldsymbol{\Sigma}$ of $z(x)$ sampled on the equispaced grids $\Omega = \{x^{(1)}, \ldots, x^{(n)}\}$ has entries $(\boldsymbol{\Sigma})_{pq} = C(|x^{(p)} - x^{(q)}|)$. Notice that the covariance matrix $\boldsymbol{\Sigma}$ is a positive semi-definite symmetric Toeplitz matrix which can be characterized by its first row $r = \mathrm{row}_1(\boldsymbol{\Sigma})$.

The key idea behind circulant embedding method is to construct a circulant matrix $\mathbf{S}$ that contains $\boldsymbol{\Sigma}$ as its upper-left submatrix. The reason for seeking a circulant embedding is the fact that, being a $m \times m$ circulant matrix, $\mathbf{S}$ has an eigendecomposition $\mathbf{S} = (1/m)\mathbf{F}\boldsymbol{\Lambda}\mathbf{F}^H$, where $\mathbf{F}$ is the standard FFT matrix of size m with entries $(\mathbf{F})_{pq} = \exp(2\pi i pq/m)$, $\mathbf{F}^H$ is the conjugate transpose of $\mathbf{F}$, and $\boldsymbol{\Lambda}$ is a diagonal matrix whose diagonal entries form the vector $\tilde{\mathbf{s}} = \mathbf{F}\mathbf{s}$ ($\mathbf{s}$ is the first row of $\mathbf{S}$).

Given a positive semi-definite circulant extension $\mathbf{S}$ of $\boldsymbol{\Sigma}$, the algorithm generates the realization of $z(x)$ sampled on Ω is shown in Table A.2. Extension to multidimensional cases can be found in [107].

Table A.1 Generating multivariate Gaussian samples by Cholesky decomposition

1:	compute the Cholesky decomposition of the positive definite symmetric covariance matrix $\boldsymbol{\Sigma} = \mathbf{L}\mathbf{L}^T$, where $\mathbf{L}$ is a lower triangular matrix
2:	generate $\mathbf{u} \sim \mathbb{N}(\mathbf{0}, \mathbf{I})$ by multiple separate calls to the scalar Gaussian generator
3:	compute $\mathbf{x} = \boldsymbol{\mu} + \mathbf{L}\mathbf{u}$ which has desired normal distributed with mean $\boldsymbol{\mu}$ and covariance matrix $\mathbf{L}\mathbb{E}[\mathbf{u}\mathbf{u}^T]\mathbf{L}^T = \mathbf{L}\mathbf{L}^T = \boldsymbol{\Sigma}$

Table A.2 Generating multivariate Gaussian samples by circulant embedding

1:	compute via the FFT the discrete Fourier transform of $\tilde{\mathbf{s}} = \mathbf{F}\mathbf{s}$ and form the vector $(\tilde{\mathbf{s}}/m)^{1/2}$
2:	generate a vector $\boldsymbol{\epsilon} = \boldsymbol{\epsilon}_1 + i\boldsymbol{\epsilon}_2$ of dimension m with $\boldsymbol{\epsilon}_1 \sim \mathbb{N}(\mathbf{0}, \mathbf{I})$ and $\boldsymbol{\epsilon}_2 \sim \mathbb{N}(\mathbf{0}, \mathbf{I})$ being independent and real random variables
3:	compute a vector $\tilde{\mathbf{e}} = \boldsymbol{\epsilon} \circ (\tilde{\mathbf{s}}/m)^{1/2}$
4:	compute via FFT the discrete Fourier transform $\mathbf{e} = \mathbf{F}\tilde{\mathbf{e}}$. The real and imaginary parts of the first n entries in $\mathbf{e}$ yield two independent realizations of $z(x)$ on Ω

References

1. Y. Xu, J. Choi, Adaptive sampling for learning Gaussian processes using mobile sensor networks. Sensors **11**(3), 3051–3066 (2011)
2. Y. Xu, J. Choi, S. Oh, Mobile sensor network navigation using Gaussian processes with truncated observations. IEEE Transactions on Robotics **27**(6), 1118–1131 (2011)
3. Y. Xu, J. Choi, S. Dass, T. Maiti, Sequential Bayesian prediction and adaptive sampling algorithms for mobile sensor networks. IEEE Trans. Autom. Control **57**(8), 2078–2084 (2012)
4. Y. Xu, J. Choi, Spatial prediction with mobile sensor networks using Gaussian processes with built-in Gaussian markov random fields. Automatica **48**(8), 1735–1740 (2012)
5. Y. Xu, J. Choi, S. Dass, T. Maiti, Efficient Bayesian spatial prediction with mobile sensor networks using Gaussian Markov random fields. Automatica **49**(12), 3520–3530 (2013)
6. M. Jadaliha, Y. Xu, J. Choi, N.S. Johnson, W. Li, Gaussian process regression for sensor networks under localization uncertainty. IEEE Trans. Signal Process. **61**(2), 223–237 (2013)
7. L.S. Muppirisetty, T. Svensson, H. Wymeersch, Spatial wireless channel prediction under location uncertainty, 4 (2015). arXiv:1501.0365
8. S. Choi, M. Jadaliha, J. Choi, S. Oh, Distributed Gaussian process regression under localization uncertainty. J. Dyn. Syst., Meas. Control **137**(3) (2015)
9. I.F. Akyildiz, W. Su, Y. Sankarasubramaniam, E. Cayirci, Wireless sensor networks: a survey. Comput. Netw. **38**(4), 393–422 (2002)
10. D. Culler, D. Estrin, M. Srivastava, Guest editors' introduction: overview of sensor networks. Computer **37**(8), 41–49 (2004)
11. P. Levis, S. Madden, J. Polastre, R. Szewczyk, K. Whitehouse, A. Woo, D. Gay, J. Hill, M. Welsh, E. Brewer et al., *TinyOS: an operating system for sensor networks, in Ambient Intelligence* (Springer, Berlin, 2005), pp. 115–148
12. C.G. Cassandras, W. Li, Sensor networks and cooperative control. Eur. J. Control **11**(4–5), 436–463 (2005)
13. S. Oh, L. Schenato, P. Chen, S. Sastry, Tracking and coordination of multiple agents using sensor networks: system design, algorithms and experiments. Proc.-IEEE **95**(1), 234–254 (2007)
14. S. Hauert, S. Leven, J. Zufferey, D. Floreano, The swarming micro air vehicle network (smavnet) project (2012)
15. P. Juang, H. Oki, Y. Wang, M. Martonosi, L.S. Peh, D. Rubenstein, Energy-efficient computing for wildlife tracking: design tradeoffs and early experiences with zebranet, in *ACM Sigplan Notices*, vol. 37, no. 10 (ACM, 2002), pp. 96–107

Y. Xu et al., *Bayesian Prediction and Adaptive Sampling Algorithms for Mobile Sensor Networks*, SpringerBriefs in Control, Automation and Robotics, DOI 10.1007/978-3-319-21921-9

16. V. Dyo, S. A. Ellwood, D. W. Macdonald, A. Markham, C. Mascolo, B. Pásztor, S. Scellato, N. Trigoni, R. Wohlers, K. Yousef, Evolution and sustainability of a wildlife monitoring sensor network, in *Proceedings of the 8th ACM Conference on Embedded Networked Sensor Systems* (ACM, 2010), pp. 127–140
17. J. Huisman, H.C.P. Matthijs, P.M. Visser, *Harmful Cyanobacteria* (Springer, New York, 2005)
18. K. Johnk, J. Huisman, J. Sharples, B. Sommeijer, P. Visser, J. Stroom, Summer heatwaves promote blooms of harmful cyanobacteria. Glob. Change Biol. **14**(3), 495–512 (2008)
19. I. Chorus, J. Bartram, *Toxic Cyanobacteria in Water: A Guide to Their Public Health Consequences, Monitoring, and Management* (Sponpress, London, 1999)
20. G.S. Sukhatme, A. Dhariwal, B. Zhang, C. Oberg, B. Stauffer, D.A. Caron, Design and development of a wireless robotic networked aquatic microbial observing system. Environ. Eng. Sci. **24**(2), 205–215 (2007)
21. Y. Wang, R. Tan, G. Xing, X. Tan, J. Wang, R. Zhou, Spatiotemporal aquatic field reconstruction using cyber-physical robotic sensor systems. ACM Trans. Sensor Netw. (TOSN) **10**(4), 57 (2014)
22. J. Lee, M. Roh, K. Kim, D. Lee, Design of autonomous.underwater vehicles for cage aquafarms, in *Proceedings of the 2007 IEEE Intelligent Vehicles Symposium*, Istanbul, Turkey, 2007, pp. 938–943
23. J. Laut, E. Henry, O. Nov, M. Porfiri, Development of a mechatronics-based citizen science platform for aquatic environmental monitoring. IEEE Trans. mechatron. **19**(5), 1541–1551 (2014)
24. I. Vasilescu, K. Kotay, D. Rus, M. Dunbabin, P. Corke, Data collection, storage, and retrieval with an underwater sensor network, in *Proceedings of the 3rd International Conference on Embedded Networked Sensor Systems* (ACM, 2005), pp. 154–165
25. A. Marino, G. Antonelli, A.P. Aguiar, A. Pascoal, S. Chiaverini, A decentralized strategy for multirobot sampling/patrolling: theory and experiments. *IEEE Transactions on Control Systems Technology* (2014)
26. F. Zhang, O. En-Nasr, E. Litchman, X. Tan, Autonomous sampling of water columns using gliding robotic fish: control algorithms and field experiments, in *Proceedings of 2015 IEEE Conference on Robotics and Automation (ICRA), Seattle*. IEE, May (2015)
27. J. Choi, D. Milutinović, Tips on stochastic optimal feedback control and Bayesian spatiotemporal models: applications to robotics, *J. Dyn. Syst., Meas. Control* **137**(3) (2015)
28. S.S. Mupparapu, S.G. Chappell, R.J. Komerska, D.R. Blidberg, R. Nitzel, C. Benton, D.O. Popa, A.C. Sanderson, Autonomous systems monitoring and control (ASMAC)—an AUV fleet controller, in *IEEE/OES Autonomous Underwater Vehicles*, 2004, pp. 119–126 (2004)
29. C.L. Nickell, C.A. Woolsey, D.J. Stilwell, A low-speed control module for a streamlined AUV, in *Proceedings of MTS/IEEE OCEANS*, Boston, 2005, pp. 1680–1685
30. P.R. Bandyopadhyay, Trends in biorobotic autonomous undersea vehicles. IEEE J. Ocean. Eng. **30**, 109–139 (2005)
31. C.C. Ericksen, T.J. Osse, R.D. Light, T. Wen, T.W. Lehman, P.L. Sabin, J.W. Ballard, A.M. Chiodi, Seaglider: a long-range autonomous underwater vehicle for oceanographic research. IEEE J. Ocean. Eng. **26**, 424–436 (2001)
32. J. Sherman, R.E. Davis, W.B. Owens, J. Valdes, The autonomous underwater glider "spray". IEEE J. Ocean. Eng. **26**, 437–446 (2001)
33. D.C. Webb, P.J. Simonetti, C.P. Jones, "SLOCUM": an underwater glider propelled by environmental energy. IEEE J. Ocean. Eng. **26**, 447–452 (2001)
34. D.L. Rudnick, C.C. Eriksen, D.M. Fratantoni, M.J. Perry, Underwater gliders for ocean research. Mar. Technol. Soc. J. **38**, 48–59 (2004)
35. E. Fiorelli, N.E. Leonard, P. Bhatta, D.A. Paley, R. Bachmayer, D.M. Fratantoni, Multi-AUV control and adaptive sampling in Monterey Bay. IEEE J. Ocean. Eng. **31**(4), 935–948 (2006)
36. N.E. Leonard, D.A. Paley, F. Lekien, R. Sepulchre, D.M. Fratantoni, R. Davis, Collective motion, sensor networks, and ocean sampling. Proc. IEEE **95**(1), 48–74 (2007)
37. N.E. Leonard, D.A. Paley, R.E. Davis, D.M. Fratantoni, F. Lekien, F. Zhang, Coordinated control of an underwater glider fleet in an adaptive ocean sampling field experiment in Monterey Bay. J. Field Robot. **27**(6), 718–740 (2010)

38. C. Zhang, A. Siranosian, M. Krstic, Extremum seeking for moderately unstable systems and for autonomous vehicle target tracking witwith position measurements. Automatica **43**(10), 1832–1839 (2007)
39. M.S. Stankovic, D.M. Stipanovic, Extremum seeking under stochastic noise and applications to mobile sensors. Automatica **46**(8), 1243–1251 (2010)
40. J. Le Ny, G.J. Pappas, Adaptive deployment of mobile robotic networks. IEEE Trans. Autom. Control **58**(3), 654–666 (2013)
41. K.M. Lynch, I.B. Schwartz, P. Yang, R.A. Freeman, Decentralized environmental modeling by mobile sensor networks. IEEE Trans. Robot. **24**(3), 710–724 (2008)
42. J. Choi, S. Oh, R. Horowitz, Distributed learning and cooperative control for multi-agent systems. Automatica **45**, 2802–2814 (2009)
43. N.A. Atanasov, J. Le Ny, G.J. Pappas, Distributed algorithms for stochastic source seeking with mobile robot networks, *J. Dyn. Syst. Meas. Control* **137**(3) (2015)
44. A. Krause, A. Singh, C. Guestrin, Near-optimal sensor placements in Gaussian processes: theory, efficient algorithms and empirical studies. J. Mach. Learn. Res. **9**, 235–284 (2008)
45. J. Cortés, Distributed kriged Kalman filter for spatial estimation. IEEE Trans. Autom. Control **54**(12), 2816–2827 (2009)
46. R. Graham, J. Cortés, Cooperative adaptive sampling of random fields with partially known covariance. Int. J. Robust Nonlinear Control **1**, 1–2 (2009)
47. R. Graham, J. Cortés, Adaptive information collection by robotic sensor networks for spatial estimation. *IEEE Trans. Autom. Control* **57**(6), 1404–1419 (2012)
48. Y. Xu, J. Choi, Stochastic adaptive sampling for mobile sensor networks using kernel regression. Int. J. Control, Autom. Syst. **10**(4), 778–786 (2012)
49. D. Varagnolo, G. Pillonetto, L. Schenato, Distributed parametric and nonparametric regression with on-line performance bounds computation. Automatica **48**(10), 2468–2481 (2012)
50. M. Jadaliha, J. Lee, J. Choi, Adaptive control of multiagent systems for finding peaks of uncertain static fields. J. Dyn. Syst. Meas. Control **134**(5) (2012)
51. C.-H. Moeng, A large-eddy-simulation model for the study of planetary boundary-layer turbulence. J. Atmos. Sci. **41**(13), 2052–2062 (1984)
52. N. Cressie, Kriging nonstationary data. J. Am. Stat. Assoc. **81**(395), 625–634 (1986)
53. C.E. Rasmussen, C.K.I. Williams, *Gaussian Processes for Machine Learning* (The MIT Press, Cambridge, 2006)
54. California partners for advanced transit and highways. http://www.path.berkeley.edu. (2015)
55. P. Seiler, A. Pant, K. Hedrick, Disturbance propagation in vehicle strings. IEEE Trans. Autom. Control **49**(10), 1835–1842 (2004)
56. B. Zhang, G. Sukhatme, Adaptive sampling for estimating a scalar field using a robotic boat and a sensor network, in *2007 IEEE International Conference on Robotics and Automation* (IEEE, 2007), pp. 3673–3680
57. M. Jadaliha, J. Choi, Environmental monitoring using autonomous aquatic robots: sampling algorithms and experiments. IEEE Trans. Control Syst. Technol. **21**(3), 899–905 (2013)
58. Y. Wang, R. Tan, G. Xing, X. Tan, J. Wang, R. Zhou, Spatiotemporal aquatic field reconstruction using robotic sensor swarm, in *2012 IEEE 33rd Real-Time Systems Symposium* (2012)
59. B. Grocholsky, J. Keller, V. Kumar, G. Pappas, Cooperative air and ground surveillance. IEEE Robot. Autom. Mag. **13**(3), 16–25 (2006)
60. C. Tomlin, G.J. Pappas, S. Sastry, Conflict resolution for air traffic management: a study in multiagent hybrid systems. Autom. Control, IEEE Trans. **43**(4), 509–521 (1998)
61. R.P. Anderson, E. Bakolas, D. Milutinović, P. Tsiotras, Optimal feedback guidance of a small aerial vehicle in a stochastic wind. J. Guid. Control Dyn. **36**(4), 975–985 (2013)
62. M.M. Zavlanos, G.J. Pappas, Dynamic assignment in distributed motion planning with local coordination. IEEE Trans. Robot. **24**(1), 232–242 (2008)
63. M.M. Zavlanos, G.J. Pappas, Distributed connectivity control of mobile networks. *IEEE Trans. Rob.* **24**(6), 1416–1428 (2008)
64. A. Jadbabaie, J. Lin, A.S. Morse, Coordination of groups of mobile autonomous agents using nearest neighbor rules. IEEE Trans. Autom. Control **48**(6), 988–1001 (2003)

65. R. Olfati-Saber, Flocking for multi-agent dynamic systems: algorithms and theory. IEEE Trans. Autom. Control **51**(3), 401–420 (2006)
66. W. Ren, R.W. Beard, Consensus seeking in multiagent systems under dynamically changing interaction topologies. IEEE Trans. Autom. Control **50**(5), 655–661 (2005)
67. M. Mesbahi, M. Egerstedt, *Graph theoretic methods in multiagent networks* (Princeton University Press, Princeton, 2010)
68. F. Bullo, J. Cortés, S. Martínez, *Distributed Control of Robotic Networks*, Applied Mathematics Series (Princeton University Press, Princeton, 2009)
69. R.M. Murray, Recent research in cooperative control of multivehicle systems. J. Dyn. Syst. Meas. Control **129**(5), 571–583 (2007)
70. Y. Cao, W. Yu, W. Ren, G. Chen, An overview of recent progress in the study of distributed multi-agent coordination. IEEE Trans. Ind. Inf. **9**(1), 427–438 (2013)
71. N. Cressie, *Statistics for Spatial Data* (A Wiley-Interscience Publication, John Wiley and Sons Inc, New York, 1991)
72. T.M. Cover, J.A. Thomas, *Elements of Information Theory*, 2nd edn. (Wiley, Hoboken, 2006)
73. A. Singh, A. Krause, C. Guestrin, W. Kaiser, Efficient informative sensing using multiple robots. J. Artif. Intell. Res. **34**(1), 707–755 (2009)
74. M. Gibbs, D.J.C. MacKay, Efficient implementation of Gaussian processes (1997) http://www.cs.toronto.edu/mackay/gpros.ps.gz
75. D.J.C. MacKay, Introduction to Gaussian processes. NATO ASI Ser. F Comput. Syst. Sci. **168**, 133–165 (1998)
76. A. Krause, C. Guestrin, A. Gupta, J. Kleinberg, Near-optimal sensor placements: maximizing information while minimizing communication cost, in *Proceedings of the 5th International Conference on Information Processing in Sensor Networks* (2006), pp. 2–10
77. J. Choi, J. Lee, S. Oh, Biologically-inspired navigation strategies for swarm intelligence using spatial Gaussian processes, in *Proceedings of the 17th International Federation of Automatic Control (IFAC) World Congress* (2008)
78. J. Choi, J. Lee, S. Oh, Swarm intelligence for achieving the global maximum using spatio-temporal Gaussian processes, in *Proceedings of the 27th American Control Conference (ACC)* (2008)
79. A.J. Smola, P. Bartlett, Sparse greedy Gaussian process regression, in *Advances in Neural Information Processing Systems, 13* (2001)
80. C.K.I. Williams, M. Seeger, Using the Nyström method to speed up kernel machines, in *Advances in Neural Information Processing Systems, 13* (2001)
81. N. Lawrence, M. Seeger, R. Herbrich, Fast sparse Gaussian process methods: the informative vector machine, in *Advances in Neural Information* (2003)
82. M. Seeger, Bayesian Gaussian process models: PAC-Bayesian generalisation error bounds and sparse approximations. Ph.D. dissertation, School of Informatics, University of Edinburgh (2003)
83. V. Tresp, A Bayesian committee machine. Neural Comput. **12**(11), 2719–2741 (2000)
84. C.M. Bishop, *Pattern Recognition and Machine Learning* (Springer, New York, 2006)
85. M. Gaudard, M. Karson, E. Linder, D. Sinha, Bayesian spatial prediction. Environmental and Ecological Statistics **6**(2), 147–171 (1999)
86. H. Rue, H. Tjelmeland, Fitting Gaussian Markov random fields to Gaussian fields. Scand. J. Stat. **29**(1), 31–49 (2002)
87. N. Cressie, N. Verzelen, Conditional-mean least-squares fitting of Gaussian Markov random fields to Gaussian fields. Comput. Stat. Data Anal. **52**(5), 2794–2807 (2008)
88. L. Hartman, O. Hössjer, Fast kriging of large data sets with Gaussian Markov random fields. Comput. Stat. Data Anal. **52**(5), 2331–2349 (2008)
89. J. Le Ny, G. Pappas, On trajectory optimization for active sensing in Gaussian process models, in *Decision and Control, 2009 Held Jointly with the 2009 28th Chinese Control Conference. CDC/CCC 2009. Proceedings of the 48th IEEE Conference on*, 2010, pp. 6286–6292
90. S.M. Kay, *Fundamentals of Statistical Signal Processing: Estimation Theory*. (Prentice Hall, Inc., Upper Saddle River, 1993)

91. J. Quiñonero-Candela, C.E. Rasmussen, A unifying view of sparse approximate Gaussian process regression. J. Mach. Learn. Res. **6**, 1939–1959 (2005)
92. H. Rue, L. Held, *Gaussian Markov Random Fields: Theory and Applications*. (Chapman & Hall, Upper Saddle River, 2005)
93. C.K.I. Williams, C.E. Rasmussen, Gaussian processes for regression. Adv. Neural Inf. Process. Syst. **8**, 514–520 (1996)
94. D.J. Nott, W.T.M. Dunsmuir, Estimation of nonstationary spatial covariance structure. Biometrika **89**(4), 819–829 (2002)
95. J.Q. Shi, T. Choi, *Gaussian Process Regression Analysis for Functional Data* (CRC Press, New York, 2011)
96. J. Nocedal, S.J. Wright, *Numerical Optimization* (Springer, Berlin, 1999)
97. W.W. Hager, H. Zhang, A survey of nonlinear conjugate gradient methods. Pac. J. Optim. **2**(1), 35–58 (2006)
98. M. Mandic, E. Franzzoli, Efficient sensor coverage for acoustic localization, in *Proceedings of the 46th IEEE Conference on Decision and Control* (2007), pp. 3597–3602
99. S. Martínez, F. Bullo, Optimal sensor placement and motion coordination for target tracking. Automatica **42**(4), 661–668 (2006)
100. F. Pukelsheimi, *Optimal Design of Experiments* (Wiley, New York, 1993)
101. A.F. Emery, A.V. Nenarokomov, Optimal experiment design. Meas. Sci. Technol. **9**(6), 864–876 (1998)
102. C.K.I. Williams, F. Vivarelli, Upper and lower bounds on the learning curve for Gaussian processes. Mach. Learn. **40**(1), 77–102 (2000)
103. P. Sollich, A. Halees, Learning curves for Gaussian process regression: approximations and bounds. Neural Comput. **14**(6), 1393–1428 (2002)
104. D.P. Bertsekas, W.W. Hager, O.L. Mangasarian, *Nonlinear Programming* (Athena Scientific, Belmont, 1999)
105. G.M. Mathews, H. Durrant-Whyte, M. Prokopenko, Decentralised decision making in heterogeneous teams using anonymous optimisation. Robot. Auton. Syst. **57**(3), 310–320 (2009)
106. W. Rudin, *Principles of Mathematical Analysis* (McGraw-Hill, New York, 1976)
107. C.R. Dietrich, G.N. Newsam, Fast and exact simulation of stationary Gaussian processes through circulant embedding of the covariance matrix. SIAM J. Sci. Comput. **18**(4), 1088–1107 (1997)
108. S. Oh, Y. Xu, J. Choi, Explorative navigation of mobile sensor networks using sparse Gaussian processes, in *Proceedings of the 49th IEEE Conference on Decision and Control (CDC)* (2010)
109. L. Devroye, *Non-uniform Random Variate Generation* (Springer, New York, 1986)
110. R. Olfati-Saber, R. Franco, E. Frazzoli, J.S. Shamma, Belief consensus and distributed hypothesis testing in sensor networks, *Networked Embedded Sensing and Control*, pp. 169–182 (2006)
111. T. Gneiting, Compactly supported correlation functions. J. Multivar. Anal. **83**(2), 493–508 (2002)
112. D.P. Bertsekas, J.N. Tsitsiklis, *Parallel and Distributed Computation: Numerical Methods* (Prentice Hall, Englewood Cliffs, 1999)
113. R. Olfati-Saber, J.A. Fax, R.M. Murray, Consensus and cooperation in networked multi-agent systems. Proceedings of the IEEE **95**(1), 215–233 (2007)
114. H. Akaike, A new look at the statistical model identification. IEEE Trans. Autom. Control **19**(6), 716–723 (1974)
115. M. Corporation, Official website of Kinect for Xbox 360. http://www.xbox.com/en-US/kinect
116. F. Lindgren, H. Rue, J. Lindström, An explicit link between Gaussian fields and Gaussian Markov random fields: the stochastic partial differential equation approach. J. R. Stat. Soc.: Ser. B **73**(4), 423–498 (2011)
117. H. Rue, S. Martino, N. Chopin, Approximate Bayesian inference for latent Gaussian models by using integrated nested Laplace approximations. J. R. Stat. Soc.: Ser. B (Stat. Methodol.) **71**(2), 319–392 (2009)
118. Y. Xu, J. Choi, S. Dass, T. Maiti, Bayesian prediction and adaptive sampling algorithms for mobile sensor networks, in *Proceedings of the 2011 American Control Conference (ACC)* (2011), pp. 4095–4200